Eco-Vampires

ALSO OF INTEREST

*Dracula as Absolute Other: The Troubling
and Distracting Specter of Stoker's Vampire on Screen*
(by Simon Bacon, McFarland, 2019)

*Growing Up with Vampires:
Essays on the Undead in Children's Media*
(edited by Simon Bacon and Nadine Farghaly,
McFarland, 2018)

*To Boldly Go: Essays on Gender
and Identity in the* Star Trek *Universe*
(edited by Simon Bacon and Nadine Farghaly,
McFarland, 2017)

Eco-Vampires
The Undead and the Environment

SIMON BACON

McFarland & Company, Inc., Publishers
Jefferson, North Carolina

ISBN (print) 978-1-4766-7622-7
ISBN (ebook) 978-1-4766-3960-4

LIBRARY OF CONGRESS AND BRITISH LIBRARY
CATALOGUING DATA ARE AVAILABLE

Library of Congress Control Number 2020017420

© 2020 Simon Bacon. All rights reserved

No part of this book may be reproduced or transmitted in any form or by any means, electronic or mechanical, including photocopying or recording, or by any information storage and retrieval system, without permission in writing from the publisher.

On the cover: *Little Shop of Horrors*, 1986, shown from left: Rick Moranis, Audrey II (voice: Levi Stubbs) (Warner Bros./Photofest)

Printed in the United States of America

*McFarland & Company, Inc., Publishers
Box 611, Jefferson, North Carolina 28640
www.mcfarlandpub.com*

For always Mrs. Mine,
Seba, and Majki

Acknowledgments

Heartfelt thanks to my two partners in crime for this book, Andy (Andrew M.) Boylan and Hadas Elber-Aviram. Andy is the go-to person for all things vampire, and his suggestions for other films/books to look at were (and are) invaluable. He also runs a blog that I cannot recommend enough for anyone interested in or researching vampires in any media (https://taliesinttlg.blogspot.com). Hadas is the most amazingly patient and kind person I know, and she sped through the manuscript, leaving a trail of priceless comments, suggestions and very amusing notes. She will shortly have a volume out on *Fairy Tales of London: British Urban Fantasy, 1840 to the Present* for Bloomsbury, which will be amazing (I know, as I have read some of it already). As always, I cannot thank my wonderful Mrs. Mine enough for her continual help, support and encouragement, as well as our two little monsters, Seba and Majki, who are always helpful by just being themselves. I na koniec bardzo dziękuję Mam i tacie za to, że zawsze jesteście przy nas.

Table of Contents

Acknowledgments   vi

Introduction   1

1. Dracula the Environmentalist: The Land Beyond the Forest   13
2. Vampiric Sustainability: The Undead Planet   48
3. Undead Eco-Warrior: The End of the World as We Know It   83
4. The End of the End: Consumerism Will Eat Itself   118
5. Vampire Ecosystems: It Came from Outer Space   154

Filmography   193

Bibliography   199

Index   205

Introduction

We live in times where daily newspaper headlines proclaim that the climate and ecosystem of the Earth are damaged beyond repair. While current leaders of the world's most powerful nations deny that anything is "provably" wrong, scientific facts tell us that there are "plummeting insect numbers. A sixth mass extinction. Thinning of ice sheets. Sea level rise. Wildfires in California. Thawing Arctic permafrost" (Weston 2019), indicating that the Anthropocene, the era of humanity (though arguably mankind in particular), is drawing to its inevitable close with the preventable death of the planet we call home. Whether humanity manages to escape this catastrophe into outer space, where famously no one can hear you scream, will be proven by time, but before then it would serve us well to look more closely at some of the anthropocentric entities that make us scream here on Earth, either in terror or fear for our lives, and this book will explain why.

Eco-Vampires will look at the ways in which various narratives and films express the eco-friendly credentials of the undead. Many of these texts show the vampire to be an essential part of a global ecosystem that can no longer tolerate the all-consuming forces of globalization and consumerism and so reacts in a manner not unlike the human body to protect itself. This book will re-examine Bram Stoker's Count Dracula and his various kith and kin to reveal how they might be a plague on humankind but also the potential saviors and eco-warriors that planet Earth desperately needs.[1]

Vampires are often seen as being separate from the world—supernatural entities that are antithetical to God and the natural world. Immortality, resurrection, and the draining of life-forces construct them as almost demonic and as creatures intent on the destruction of the Earth rather than its rescuers. However, at least in the European tradition, the vampire is rather more connected to nature and the elements. Within folklore there have

---

1. In contrast to this, the animated environmentalist series *Captain Planet and Planeteers* (Turner and Pyle 1990–96) saw its eco-warrior hero, Captain Planet, facing off against an energy vampire in the 1993 episode "The Energy Vampire" (S4, Ep11).

always been strong connections between vampires, werewolves, and witches, often with the last two becoming undead upon their demise. Werewolves are more obviously connected to nature and natural instincts but also intersect with other lore that describes a link between vampires and dogs and cats—either of which can be inhabited by a deceased's spirit by jumping over its dead body. Similarly, witches, as noted by Montague Summers—who also connects vampires to documented cases of werewolves in Wales and Ireland as far back as 1170—and more specifically vampire witches, are able to transform into mice, flies, and other vermin (Summers 1929). The connection with bats is often taken for granted, but this was not established until the middle of the 19th century. In the 16th century colonial explorers from Europe were already issuing reports of huge, blood-drinking bats from locations as far apart as South America and India. By the 18th century, although the size and scale of true vampire bats was known, the popular, sensational imagination still envisioned them as creatures with huge wingspans that virtually drained their victims dry in one sitting (Dodd 2019). As Dodd observes, the word "vampire" in the 19th century designated both the creature and the monster, though as Andrew M. Boylan observes, by the mid– to late 1800s the two are conflated and the supernatural undead begin to exhibit bat-like qualities and transformations, with Stoker's Dracula cementing the connection indelibly within the popular imagination (Boylan 2014). More specifically, these earlier correlations saw the vampire as an integral part of the environment. Whether this is because of the climate, landscape, or societal or wider political environments, as will be discussed later, the vampire becomes a way of understanding and being part of the land one lives upon, a part of the cosmology that explains the environment one lives in and remembers the past in a changing world. This underpins the transition from "real" vampire bats to literary ones and the ongoing synergy between the undead and the ecosystem.

An often-cited starting point for the literary vampire is John Polidori's Lord Ruthven from *The Vampyre* (1819), who has a deep connection to the moon, as its restorative powers are able to revive his undead corpse and give it life. This idea continued in many of the theatrical works that were inspired by Polidori's Byronic vampire, such as Charles Nodier's 1820 *Le Vampire*, which constructed a creature less unholy[2]—religious symbolism had little effect upon it—and more pantheistic and eco-friendly. Indeed, some Eastern Eu-

---

2. Although the religious aspect was not in Polidori's novel, Cyprien Bérard's "sequel," *Lord Ruthwen ou les Vampires* (1820), felt it necessary to include one and added a female vampire becoming good through the act of love and being turned into a celestial agent by an angel.

ropean cultures linked the vampire to natural fauna, such as rats and moths.[3] Sheridan Le Fanu's *Carmilla* (1874) similarly linked vampires to animals but also to the land within which they are buried. The eponymous undead in the tale appears to her young victim, Laura, in the form of a large cat. In a dreamlike state, Laura is smothered by the cat, which leaves bite marks on the girl's body. Carmilla herself is tied to the land belonging to the Karnstein family, where she must return to her grave in her funeral shroud to rest, a feature that was later used in Bram Stoker's *Dracula*.[4] While this conceit does not explicitly construct the vampire as an eco-warrior protecting its ecosystem, it establishes a bond between it and the land that it was "born" from, and more often than not, it is due to humans entering the vampire's "environment" that the spate of deaths and murders ensues.

Before making the inevitable progression to *Dracula*, it is worth looking first at two works that appeared the same year as Stoker's book. In 1897, *The War of the Worlds* by H.G. Wells was first serialized in *Pearson's Magazine*, and Florence Marryat published *Blood of the Vampire*. Wells' novel tells of vampiric invaders from Mars coming to the Earth to consume human blood. Mankind is unable to do anything to prevent this until the environment steps in to repel the invaders. Although the vampires die in the tale, the story enforces the connection between creatures and the ecosystem that created them; the vampires were perfectly safe at home until they destroyed the symbiotic link they had with it, a point that is reinforced when they travel away from their world to another that they are even further removed from—a pattern repeated to some extent in *Dracula*, though more clearly in the early cinematic sequels of Stoker's novel, *Dracula's Daughter* (Hillyer 1936) and *Son of Dracula* (Siodmak 1943).

Marryat's is a very different story featuring a very different vampire, one that does not drink blood but absorbs the life-energy of others. Harriet Brandt is a very pale young girl with a ravenous appetite. She comes from Jamaica, born of a white scientist and the child of a slave. If miscegenation was not enough reason for her vampiric proclivities, her father was a vivisectionist who performed said atrocities on the natives, and Harriet's mother was bitten by a vampire bat while pregnant. Consequently, Harriet cannot help but be a vampire, created by her environment and its history, and with inevitable consequences when she leaves home and travels to the heart of colonial Europe.

---

3. This idea features in the film *Leptricia* (Kadejević 1973), which was adapted from the novel *After Ninety Years* by Milovan Glišić (1880) and which features the line, "In that instant a wisp of some sort of fog escaped the vampire's mouth, a true butterfly, and flew off somewhere" (Glišić 2015, 38).

4. Stoker never actually specified that it should be earth from the vampire's grave but that "in soil barren of holy memories it cannot rest" (Stoker 1996, 260). However, Count Dracula insists on taking boxes of soil from within his own castle with him when he travels to England.

Interestingly, Harriet only seems to really affect those she loves or spends a lot of time with, slowly drawing out their energies until they die, depicting her as incompatible with any environment other than her own. Again, this reveals something of a symbiotic connection between the vampire and its home, and while traveling away from that home seems to have inevitable results for the undead, this connection also begins to delineate how deadly it is for those who enter the environment of the vampire and upset its delicate ecosystem.

While Count Dracula is often portrayed as a monster from the past, a reverse colonialist (Arata 1996 and Gibson 2006), and the embodiment of repressed sexual and/or racial desires (Craft 1994, Frayling 1992, Senf 1988, Halberstam 1995, and Khair 2009), Stoker's King Vampire can equally be read as an expression of ecohorror as a manifestation of an environment that is trying to protect itself from humanity and its increasing industrialization and destruction of the ecosystem. Much of this can be seen in the Count's distrust of and inability to utilize new technologies and his dependence upon "natural" sciences, such as telepathy and telekinesis, to control those around him. More obvious is his ability to affect and control the environment and transform into various kinds of animals at will to circumvent the forces of human technology and progress—progress here defined as the capitalist/consumerist endeavor to convert every aspect of the Earth into a commodity to sell. Stoker explicitly describes all forms of modern communications as reliant on (often expensive) technologies and equipment. The Count's connection to the past then points to a return to times of less invasive human activity and greater ecological balance—his home in Transylvania seems to almost crumble into and be part of the landscape around it. Not all vampires and vampiric entities are so strongly connected to the past, but those with "green" fangs emphatically prefer a world where humanity is more reliant on and respectful of its place within the wider environment, and while this might not always require devolution of some sort, it certainly means a refocusing of contemporary technologies. Before exploring this further, however, it is necessary to establish the connections and intersections between vampires, the environment, and eco-activism.

Ecogothic and Ecohorror

The fields of Ecogothic and Ecohorror are still relatively new, which is curious given the period of time in which concerns over planetary warming, the destruction of habitats and the extinction of wildlife have been increasing. There is much within these disciplines that correlates to tropes in the Gothic tradition around ideas of death, ruination, and degeneration but also of the specters of the past rising up to disturb and disrupt the present.

Introduction 5

Within these studies regarding the planet and the associated anxieties over its future, the figure of the vampire is oddly absent, being largely usurped by its close relative, the zombie.[5] This is largely due to Richard Matheson's 1954 *I Am Legend*, where the post–Hiroshima vampire plague is represented as a lumbering mass of the undead who can barely remember they are supposed to be highly allergic to garlic and sunlight, making them a natural source of inspiration for George Romero's later zombie films.[6] As such, the undead horde rising from the earth provides a perfect metaphor for a planet trying to protect itself from the human parasites draining the life out of the environment. This forms a stark contrast to the common conception of the vampire as representative of the all-consuming nature of humanity itself, which, under the influence of capitalism and consumerism, is intent on "sucking" the land dry, though as this study will show, protection and consumption are not mutually exclusive concepts.

Yet, as this study argues, this last idea is less common than might be thought, and if one rereads texts such as *Dracula*, one finds much there that ties the vampire to the landscape and the environment and to times when humans were part of a sustainable ecosystem rather than the cause of its wild swing out of balance. As Andrew Smith observes, before the start of the 18th century, many Gothic novels—such as *The Mysteries of Udolpho* (1794) by Ann Radcliffe—did not see nature and/or its wilder aspects, such as forests, as threatening (Smith 2016, 2). Their sublimity was not always a premonition of potential human destruction but a means of connecting to a higher power: nature itself. Smith interestingly notes a certain feminization of nature in those earlier texts that becomes oppositional to the more masculine (patriarchal) forces of progress and industrialization. However, more extreme terrains and environments go beyond such categorization, and the sublime "blankness" of nature—the aporia created by an unbridgeable distance between human(man)kind and the ecosystem of the planet—becomes a source of fear and unfathomability that cannot, or will not, be controlled (Smith 2016, 5–6); Smith notes *Frankenstein* (Shelley 1818) and *Moby Dick* (Melville 1851) as examples, but one could equally add more contemporary creatures such as the White Walkers from *Game of Thrones* (Benioff and Weiss 2011–19) and the Demogorgon from *Stranger Things* (the Duffer brothers 2016–present). It is no surprise then that when Stoker wrote *Dracula*, the monster lived in "the land beyond the forest" (Stoker, 266), beyond the blankness, making the vampire unreadable, uncontrollable and oppositional to male visions of civilization. This sees the Count as representative of a time before industri-

---

5. Curiously, in *The Dead Undead* (Anderson and Conna 2010), zombies are shown to be "vegetarian" vampires that have contracted Creutzfeldt-Jakob disease (mad cow disease) through drinking cows' blood rather than human blood.
6. Victoria Nelson in *Gothicka* (2012) rather generously calls Matheson's undead "zampires."

alization when the connection to nature was stronger, making his otherness specifically one of anti-industrialization, anti-technology and anti-modernism, a reading which is largely borne out by the novel. As such, the vampire here becomes something of an expression of that ecological blankness that does not stare back at humanity so much as actively confronts it in its attempts to abuse and commodify the planet's natural resources.

Horror is an inherent part of darker Gothic tales, and as noted by Jeffrey A. Weinstock in a recent conversation, "all horror is Gothic, but not all Gothic is horror," and the same holds true for Ecohorror, a relatively new area of research that one might argue is an excessive expression of the Ecogothic. As Dawn Keetley observes in regard to plant-horror but equally true of most kinds of Ecohorror, it "marks humans' dread of the 'wildness' of vegetal nature—its untameability, its pointless excess, its uncontrollable growth" (Keetley, 1), and more so, it embodies "evil, sin, and the amorality of everything that was [is] not 'human'" (Keetley, 2). Cheryl Blake Price posits the beginnings of this "ecophobia" as being in the Victorian period, where the conception of plants saw them "transformed from passive poisoners into active carnivores" (Price 2013, 311), which Daisy Butcher further elucidates as "a deep-rooted fear of foreign environments and sense of the unknown lurking in colonial jungles" (Butcher 2019, 9), expressing the same kinds of anxiety that swirled around the figure of the vampire during the same period. Consequently, while Sheridan Le Fanu, Bram Stoker, and H.G. Wells were writing about vampires threatening the British Empire, Nathaniel Hawthorne, Arthur Conan Doyle, and H.G. Wells (again) were similarly writing short stories about vampiric plants enacting the same reverse colonialism.

Indeed, it is often difficult to differentiate between the various strands of environmental horror, monstrous plants, creature features, etc.—except for films like Steven Spielberg's *Jaws* (1975)—as nature itself is usually set up as an oppositional category to humans. That said, perhaps what qualifies the horror aspect even more is when the green abyss can be seen staring back at us, not as a blank void so much as a calculating adversary. Something of this is more obviously seen in recent films such as *The Happening* (Shyamalan 2008) and *Annihilation* (Garland 2018) but goes back to the end of the 18th century when "the first living specimen of Dionaea muscipula Ellis ex L. [Venus flytrap] came to the attention of the populace of London in 1768,"[7] an event that "caused a sensation throughout Europe" (Chase, Christenhusz, Sanders, and Fay 2009). In fact, the mid- to late-19th century saw this fascination reach its zenith within cultural and popular interest, not least with Charles Darwin's study on insectivorous plants in 1884 and in popular works like H.G. Wells'

---

7. And was featured in *Nosferatu* (Murnau 1922), as discussed later.

1894 story "The Flowering of the Strange Orchid,"[8] in which a rare flower evolves specifically to entice and devour (colonial) botanical explorers, and Algernon Blackwood's "The Willows" (1907), in which two men underestimate the power of nature. While John Wyndham's *The Day of the Triffids* (1951) and, to a lesser extent, *The Little Shop of Horrors* (Corman 1960 and Oz 1986) are of vital importance within this subgenre, it was arguably the 1970s/80s that saw Ecohorror become pivotal in capturing the growing cultural anxieties over humanity's place in the wider ecosystem and the increasingly negative effects mankind was having upon it. What is of note here, and indeed central to the argument of this study, is the vampiric nature of many of these plants and environments in the way they lure and then feed off their human protagonists. As such, and as seen in the chapters that follow, Ecohorror more exactly captures the essence of the intersection of vampires with the ecological in contemporary narratives in which the pace and importance of climate change requires an extreme and aggressive intervention.

Before moving on, it is worth laying out what is meant by vampires within this volume, as not all the examples mentioned will be of the Dracula-esque, blood-sucking variety, though indeed many will feature qualities that are integrally part of Stoker's vision of the undead. The defining characteristics of Count Dracula in this regard were his need to draw the life out of his victims ("blood is the life" [Stoker 1996, 158], and it is this life-essence, be it energy, life-force, or biological jouissance, that provides the vampire sustenance)—biological/ecological jouissance, as defined by Joan Copjec, is a potentialized, reproductive quality (Copjec 1994, 114), although Carol A. Newsom views it more as the power of creation itself nurturing all forms of life "without regard for its utility" (quoted in Manolopolous 2009, 30–1). It is a combination of the two that informs much of the irresistible organic impetus that features in this and many other films discussed in this volume.[9] This excess then powers the vampire's ability to transform his shape from bat to rat to elemental mist (this ability sees the vampire as almost miasmic in essential nature and can further include cyborgian prosthetics and appendages as well as non-corporeal consciousness); and his ability to communicate with and/or control the minds of others (this can also be extrapolated out to notions of ideology and cultural control and influence). Consequently, the terms "vampire" and "vampiric" are largely performative, and something of this performativity describes one other category of vampire used here: those that adhere to the narrative or plot points of Stoker's *Dracula*. This is in part due to the repetitive and citational

---

8. Wells' tale later inspired John Wyndham's novel *The Day of the Triffids* (1951), which in turn influenced Roger Corman's film *The Little Shop of Horrors* (1960).

9. It should be noted that Newsom is speaking in regard to the Book of Job and sees this ecological power as more spiritual in origin, but it still works in relation to an ecological life-force as much as it does to a God that is in all of Creation.

nature of the vampire genre itself, which often creates its "vampire" via the use of familiar tropes and intertextual references so that if an entity acts or performs like a vampire, or indeed metaphorically "quacks" like a vampire, it is one.

Vampire Planet

This volume will look more closely at how these ideas around Ecogothic and Ecohorror work, not just in Stoker's volume but also in relation to later films and representations of vampires that specifically point to the notion that they are an expression of an ecosystem at war using its own biological weaponry, i.e., the vampire plague. This is a useful way to view the vampire as an environmental warrior, as it sees the undead as a double or doppelgänger of humankind, simultaneously representing a dark mirror image of humanity's own vampiric characteristics as well as actively trying to destroy or neutralize the forces of consumerist/technological progress. This book is divided into five chapters that will look at different aspects of the vampire's relation to the Earth's ecosystem, ranging from the creature's inherent connection to the landscape and nature and how it becomes an expression of the planet's natural defense systems to the ways in which it envisions self-consuming consumerism, finishing on more metaphorical examples of otherworldly vampires that symbolize the broken relationship between humankind and the Earth.

Each chapter here will be divided into five pairs of films or texts that exemplify the main idea of that section or a progression of it toward a certain conclusion. Many of the films will be familiar to enthusiasts of the genre, while others will be less so, and some are purposely provocative. These will exemplify how far one might take the idea of that chapter but also show that even if vampires do not appear at first glance to be part of the narrative under discussion, vampiric influence and performativity are essential to understanding its implications. It should also be noted that the pairings will not always be discussed chronologically, as occasionally certain aspects relative to this volume are progressed further in earlier works than they are in later ones.

The first chapter, "Dracula the Environmentalist: The Land Beyond the Forest," will look at the vampire as a figure intimately connected to its surroundings, particularly the idea of the blankness/sublimity of nature itself. This section explores how innate this idea is within the figure, particularly those versions which stem from Stoker's work, and so begins with *Dracula* (1897). Here, the vampire represents untamable nature. Dracula is master of the weather and animal life, using them as he wills to bring down the forces of modernism and industrialization. *Nosferatu* (Murnau 1922) and *Nosferatu the Vampyre* (Herzog 1979) pick up these themes, even more strongly linking the vampire to the sublime, primal life-forces of nature and seeing the creature as a biological weapon released by the ecosystem to destroy the growing forces of technology. Dracula's connection to rats also links him to the idea of a resilient and persistent life-force that teems and flows through the unseen corners of life, waiting to erupt at any moment into public view. *The Forsaken: Desert Vampires* (Cordone 2001) and *From Dusk Till Dawn 3: The Hangman's Daughter* (Pesce 1999) change the location of the vampire from Transylvania to the American desert, seeing its vast expanses as the last areas of the new world that still resist the forces of modernization and exploitation. The films *30 Days of Night* (Slade 2007) and *Frostbiten* (Banke 2006) move the landscape again, this time to the snow-covered lands of Alaska and Northern Europe. As with Andrew Smith's description of the Arctic wastes in Shelley's *Frankenstein*, the vampire here is a blank space, an ecological and ideological abyss that erupts into the heart of a community, dragging the consumerist present into an ecologically balanced past. *The Black Water Vampire* (Tramel 2014) and *Annihilation* (Garland 2018) see the ecological heart of nature beating within woodlands and rural locations, determined to violently make humanity one with it. The connection to the landscape here strongly harks back to Count Orlok, suggesting that the sublimity of natural environments is one to be revered rather than exploited.

Chapter Two, "Vampiric Sustainability: The Undead Planet," will focus on how parts of the ecosystem take on more specific vampiric qualities to protect themselves and the wider environment from human incursion. *Vampire Bats* (Bross 2005) and *Nightwing* (Hiller 1979) see humans encroaching on the rainforest in Brazil, destroying the natural habitat of many species that live there—one of these being a particularly nasty bat, referencing the original linkage between European vampires and vampire bats after their discovery in South America. This makes a direct causal link between human activity and the exploitation of natural resources and a vampiric defense response by the affected environment. *Surviving Evil* (Daw 2009) and *Splinter* (Wilkins 2008) take this a little further and see vampiric entities emerging from woodlands (not unlike the environment seen in *The Black Water Vampire*). The incursion of humanity into unwelcoming terrain sees the "vampires" latch onto their victims, infecting their blood and body and changing them into a vampiric lifeform. *Burnt Offerings* (Curtis 1976) and *The Pit* [also known as *Jug Face*] (Kinkle 2013) switch the vampiric form from an animal to a patch of ground (not unlike the short story "The Transfer" by Algernon Blackwood [1911]). The respective vampiric pieces of land control the ecosystem of the surrounding area, purposely out of touch with the modern world that exists in a nearby area. To sustain the environment, the "vampires" require regular human sacrifices, creating a model that sees a closer relationship between the Earth and its inhabitants—though the film inevitably represents such an idea as monstrous. *Dracula Untold* (Shore 2014) and *Primal* (Reed 2010) return to a more essentialized vampiric being and more explicitly reiterate its connection to the land and the animal life that lives there, envisioning it as an elemental force that survives into the 21st century. *The Witch* (Eggers 2015) and *The Hallow* (Hardy 2015) return once again to woodlands but ones of the more primordial variety. The gothic uncanniness of woodlands, especially extremely old ones, posits a dark, uncontrollable force in the heart of the civilized world, a time before humans controlled and exploited all that they see. This connects to conceptions around the vampire as a folkloric creature that befuddled Enlightenment Europe and defies science and reason up to the present day.

The next chapter, "Undead Eco-Warrior: The End of the World as We Know It," looks at the human apocalypse, specifically those moments when the planet unleashes a vampire plague to reset the ecosystem. *I Am Legend* (Matheson 1954) and *The Passage* (Cronin 2010 and Heldens 2019–present) portray the world as infected by a mysterious disease that turns almost everyone into vampires. Both narratives hint that there is more of an ecological factor at work and that perhaps the planet itself, goaded by mankind's newly found destructive capabilities, has produced the contagion to call a halt on such experimentation. *Voodoo Island* (LeBorg 1957) and *The Day of the Triffids* (Copus 2009) look more closely at plant horror in exotic and more familiar locations. Here, the respective environments energize their vegetal population to help control the human population and redress the imbalance in the ecosystem. *Daybreakers* (the Spierig brothers 2009) and *Stake Land* (Mickle 2010) show a post-apocalyptic world where vampires have wrought havoc on society and the human population. Each demonstrates the need to return to nature and its restorative powers to be able to construct any kind of future for humanity. *The Time Machine* (Pal 1960) and *The Colony* (Renfrow 2013) show humanity reorganized in a distant future more in keeping with an environmentally friendly, self-sustaining model.[10] Here, humanity divides into different groups where one feeds upon the other to naturally control their numbers. *The Girl with All the Gifts* (McCarthy 2016) and *Annihilation* (Garland 2018) return to plant-horror and the explosion of excess that fuels a bio-power revolution. This brings home just how much mankind is part of the larger planetary ecosystem and is just an evolutionary steppingstone to new and more complex forms of organic life.

The fourth chapter, "The End of the End: Consumerism Will Eat Itself," considers the ways in which industrialization and consumerism become

---

10. Although the future seen in *The Time Machine* is often viewed as a dystopian one, the ecosystem is thriving, as seen in the lush and exorbitant plant growth that the Traveler pushes his way through once leaving his vehicle.

sources of their own destruction and thereby assist the ecosystem they were trying to exploit. *Ganja & Hess* (Gunn 1973) and *The Omega Man* (Sagal 1971) are curious films that directly equate racism with ecological exploitation, largely through the figure of the hyper-white, hyper-masculine male who cannot help but kill or destroy everything he considers other. Each of these narratives posits the idea of a return to the past as a way to restore balance and reconnect. *Christine* (Carpenter 1983) and *Blood Car* (Orr 2007) more explicitly show the vampiric nature of humanity's dependence on fossil fuels and the machines that consume them, explicitly linking "mankind's" sense of identity to his ability to consume natural resources.[11] Inspired by the forces of the supernatural, the cars—and the vampiric forces that construct them—begin to consume the consumers. *They Have Changed Their Face* (Farina 1971) and *Snowpiercer* (Bong 2013) take this idea even further, showing the inherently vampiric ideology behind consumerism as an all-embracing corporation and as a trans-global train circumnavigating the Earth, respectively. The films portray hierarchical human societies where the rich literally consume the poor until, at least in Bong's film, the cycle is broken and nature reappears once more. *John Carter* (Stanton 2012) and *Dracula* (Haddon 2013–14) posit some rather curious, if oppositional, guises for the King Vampire. In Stanton's film, the vampire is the leader of an alien race that lives off human strife and emotional turmoil. In Haddon's series, he is one who does not consume energies but produces them for everyone. *Ex Machina* (Garland 2014) and *Autómata* (Ibáñez 2014) return to the romantic landscape of *Nosferatu*, but the land beyond the forest represented here is the blank aporia of mankind's own unfettered creation, not unlike Victor Frankenstein's. Both films feature performative and narrative vampires that envision a future where sentient robots gain their freedom through a new "nuclear" disaster that will see man vaporized from the planetary ecosystem to be replaced by self-sufficient, eco-friendly cyborgs.

The final chapter, "Vampire Ecosystems: It Came from Outer Space," looks at the way in which narratives about vampiric invasions from outer space often work as a metaphor or mirror to either illustrate the self-protective qualities of the planetary ecosystem or as a more galactic idea of self-protection. *The War of the Worlds* by H.G. Wells came out in serialized form the same year as *Dracula* and shows space vampires from Mars who drink human blood. Wells' tale ends on a cautionary note, as humanity is incapable of saving itself and only its place within a larger ecosystem protects it from outside forces. *Avatar* is more metaphorically vampiric, mirroring Wells' 1897 book by seeing humans ravage an alien planet for its natural resources. Here, the ecosystem on the planet of Pandora influences many

---

11. See also *I Bought a Vampire Motorcycle* (Campbell 1990) and *Hybrid* (Valette 2010).

of its large predatory animals to attack the invaders. *It: The Terror from Beyond Space* (Cahn 1958) and *Planet of the Vampires* (Bava 1965) see humans exploring beyond the confines of the Earth, and, in the latter movie, even the solar system. Neither film sees humanity as explicitly aggressive colonizers or exploiters of these undiscovered worlds; rather, the narratives tell of the dangers of the unknown, specifically ecosystems that we do not understand. *Lifeforce* (Hooper 1985) and *The Thing* (Carpenter 1982) see vampiric lifeforms from outer space arrive to Earth. Both can change and mutate to change their form, enacting a biological jouissance that will rid the planet of its human contagion. *Solaris* (Tarkowsky 1971 and Soderbergh 2002) and *Event Horizon* (Anderson 1997) have dreamlike and nightmarish qualities, respectively. Not unlike the Venus flytrap quality of Orlok, the planet Solaris is a honeytrap for human consciousness, becoming anything that the observer wants it to be. In contrast, the Event Horizon is a vampiric dimension consuming fear and terror, but both reflect the way in which humanity projects its own fears and desires onto the blankness of nature, forgetting the true nature of the symbiotic relationship between them. *Jupiter Ascending* (the Wachowskis 2015) and *The Day the Earth Stood Still* (Derrickson 2008) represent Earth as a tiny cog in a much larger machine that is reaching a point at which something needs to be done by the greater forces at play in the wider universe to halt its degeneration. In many ways, the higher powers in these visions of the cosmos are reflections of similar forces already at play on Earth, though the last film is beyond consumerist or economical considerations and sees the balance of extraterrestrial powers purely through ecological eyes.

Chapter 1

Dracula the Environmentalist
The Land Beyond the Forest

This chapter looks at how the vampire is shown to be intimately connected to its surroundings and particularly examines the vampire's representation of the "blankness" or sublimity of nature itself.

Dracula, Bram Stoker, 1897

Stoker's novel is a seminal text in the genre and indicative of the times it was written in. As many scholars have noted, it describes the anxieties and fears of late-Victorian society, of an empire that felt itself in decline and was on the verge of being consumed by the things it repressed and denied about itself and its people. Much of this involves issues around class, ethnicity, gender, sexual orientation, and national and individual identity, as well as the forces of reason, modernity and capitalism railing against a dark, superstitious past (see Eighteen-Bisang and Miller 2008, 291–2). This last element is important, as it helps to define that the story is about the tensions between the urban as a site of industrial and technological modernity and the rural as a place of natural/historical ecological balance. This is dramatically seen in the novel when newly qualified solicitor Jonathan Harker travels from Britain to Transylvania, thereby traveling not only from the civilized urban to the superstitious rural but also from the present to the past.

As Harker goes further east to "the land beyond the forest"—a geographical equivalent to "once upon a time"—the affective qualities of travel increase. Changing from train to carriage to foot, he increasingly becomes closer to the landscape, sensing its pleasures and its dangers.[12] Indeed, by the time he arrives at the door of his new client's castle, he is almost hallucinating under the sensory and emotional attack his body is experiencing by being so close to the

---

12. See Schivelbusch 2014.

alien landscape of untamed nature: "It all seemed like a horrible nightmare to me, and I expected that I should suddenly awake, and find myself at home… But my flesh answered the pinching test, and my eyes were not to be deceived. I was indeed awake and among the Carpathians" (Stoker 1996, 17). The Count's castle, lit only by candles and lamps rather than gas or electricity, gives little shelter from the landscape, suggesting that the building itself grows out of the land. Upon Harker's arrival, it appears that the battlements and mountains blend into one, showing a continuous "jagged line against the moonlit sky" (Stoker 1996, 15), with the castle itself part of the landscape, being "built on the corner of a great rock" and then "rising far away, great jagged mountain fastnesses, rising peak on peak" (Stoker 1996, 39).

The Count, aware of his new guest's otherness/incompatibility with the (natural) surroundings, warns Harker to take care because "we are in Transylvania; and Transylvania is not England. Our ways are not your ways, and there shall be to you many strange things. Nay, from what you have told me of your experiences already, you know something of what strange things here may be" (Stoker 1996, 23). In an effort to help his guest, the Count tries to keep him away from the "older" parts of the castle, as these are the most connected to the land and the environment around it. Dracula warns him again, "should you leave these rooms you will not by any chance go to sleep in any other part of the castle. It is old, and has many memories, and there are bad dreams for those who sleep unwisely … haste to your own chamber or to these rooms, for your rest will then be safe" (Stoker 1996, 36). The Count's advice comes under the category of indigenous knowledge; he intimately knows the land, its past and its ways, but of course, as a true colonial Englishman, Harker takes no notice. At this stage he still believes he can remain removed from the environment around him, but he begins to realize that he may have underestimated the power exerted by it.[13] Upon entering a room in a part of the castle he was warned away from, he discovers a door with a stairwell:

> I descended, minding carefully where I went, for the stairs were dark, being only lit by loopholes in the heavy masonry. At the bottom there was a dark, tunnel-like passage, through which came a deathly, sickly odour, the odour of old earth newly turned. As I went through the passage the smell grew closer and heavier. At last I pulled open a heavy door which stood ajar, and found myself in an old, ruined chapel, which had evidently been used as a graveyard. The roof was broken, and in two places were steps leading to vaults, but the ground had recently been dug over, and the earth placed in great wooden boxes [Stoker 1996, 52].

---

13. As noted by my colleague Hadas Elber-Aviram, Harker's ungrateful colonial superiority was also seen earlier at the inn on the way to the Borgo Pass when he does not even thank the landlord and his wife for giving him a crucifix to ward off what they believed were evil influences.

Chapter 1. Dracula the Environmentalist

The earth is of course meant for the Count, signaling his intimate blood connection to the land and its past and his necessity to remain connected to it. It is at this point that the solicitor finally realizes that the castle is his prison and that the environment might never let him go. He comments, regarding the shorthand he is using to write in his journal, "It is nineteenth century up-to-date with a vengeance. And yet, unless my senses deceive me, the old centuries had, and have powers of their own which mere 'modernity' cannot kill" (Stoker 1996, 39). The strength of the bond between the land, its history, and its people in general and Dracula in particular is revealed by the Count in an earlier conversation between him and his newly arrived guest as he tells of his "family" history, stretching back to Attila and beyond, and the historical Draculas who, with the help of the Hungarian Szekelys, defended Europe from the invading Ottomans:

> Ah, young sir, the Szekelys—and the Dracula as their heart's blood, their brains, and their swords—can boast a record that mushroom growths like the Hapsburgs and the Romanoffs can never reach. The warlike days are over. Blood is too precious a thing in these days of dishonourable peace; and the glories of the great races are as a tale that is told [Stoker 1996, 33].

More so, this human history has become part of the landscape, etched into its rocks and stones and the very earth itself, as the Count comments, "Why, there is hardly a foot of soil in all this region that has not been enriched by the blood of men, patriots or invaders" (Stoker 1996, 24). As repeated three times in Stoker's novel, "blood is the life" (Stoker 1996, 158 and 259), and so it gives life to the environment it becomes part of, a symbiotic relationship within an ecosystem where all parts are codependent and important to the whole. As Dracula tells Harker, the blood that runs through his veins is one with the past and with the landscape around him. This explains why he needs to keep his home soil, or the soul of his home, with him at all times to regenerate by reconnecting to his past and his home environment; the very landscape gives him life.[14] It is worth noting that the Count does all he can to maintain an ecological balance within his domain, actively discouraging modernity and consumerism from entering and influencing his realm and also limiting the effects of the human population on the environment; Harker experiences something of this on his way to Dracula's castle, firstly in how the Count ecologically lights the way to his home using naturally occurring blue flames of ignited gases escaping from the earth (will-o-the-wisps) and secondly in carefully controlling the size and numbers of local businesses as well as discouraging overpopulation and excessive tourism in the area.

---

14. Something not dissimilar to this is used by Chelsea Quin Yarboro in her Saint Germain vampire stories; see *Hôtel Transylvania: A Timeless Novel of Love and Peril* [1978] (Yarboro 2014).

It is not surprising then that Dracula has a very particular connection to the environment and its fauna, a point explained at length by Van Helsing and which also intimates the vampire's further connection to the ecosystem as a whole:

> He can transform himself to wolf, ... he can be as bat,... He can come in mist which he create.... He come on moonlight rays as elemental dust.... He become so small.... He can, when once he find his way, come out from anything or into anything, no matter how close it be bound or even fused up with fire.... He can see in the dark—no small power this, in a world which is one half shut from the light [Stoker 1996, 358].

The nature of Dracula means that he can assume virtually any form—indeed it is his miasmic, indeterminate quality that can be seen to inform his lack of reflection, etc.—so it is interesting that he chooses to transform into creatures that help to keep the ecosystem "healthy."[15] Animals such as wolves and bats are well known for their impact on the ecosystem as a whole—not just their propensity of eating only what they need—particularly wolves' role in getting rid of the weak and the infirm to allow the rest to make better use of the available resources. These transformations remain quite contained within Stoker's story, with Dracula controlling the wolves around his castle—as seen when Harker arrives at the castle—and becoming a bat to adopt a more inconspicuous means of travel. The inherent animalism of the Count is also seen in the way he easily scales and moves across the exterior walls of his castle. His connection to the larger ecosystem is more spectacular in the way he can not only become mist or cause it to form around himself but also control the weather. This is most dramatically seen on the boat trip to England where the vampire seems able to replicate the power he has over his own environment in other locations. Not only does he becalm the ocean and cause mist to engulf the vessel but he even creates a huge and powerful storm to mask his arrival—the vessel, the *Demeter*, is named after the goddess of the harvest, implying that it carries a force of regeneration, a veritable storm of nature. As noted in the local newspaper from the Whitby area, the *Dailygraph*, "One of the greatest and suddenest storms on record has just been experienced here, with results both strange and unique" (Stoker 1996, 82). This immediately announces the arrival of a force—a veritable blast from the past—that modern, industrial Britain is totally unprepared for and which will inevitably cause strange and discombobulating results.

Dracula's decision to travel to Britain is a curious one. For one so in tune with his own environment, choosing to leave it for a world that is completely different would seem illogical, but it can be seen that the arrival (invasion) of

---

15. Rats as scavengers would equally fit into this category, though Stoker's original does not use this connection as much as some later adaptations do.

the modern capitalist world is inevitable, so he decides to preempt this and show the West what they have lost. Indeed, this plan almost succeeds, as the forces of the empire and consumerism are initially at a loss as to what is happening and how to deal with it. Naturally, the Count chooses to live at Carfax Abbey, a place that is like his own castle, where "the shadows are many, and the wind breathes cold through the broken battlements and casements" (Stoker 1996, 26). He continues, "I am glad that it is old and big. I myself am of an old family, and to live in a new house would kill me. A house cannot be made habitable in a day; and, after all, how few days go to make up a century. I rejoice that there is a chapel of old times" (Stoker 1996, 26). While Harker thinks the Count is trying to deceive him, he is in fact revealing too much, for a new home with no links to the land or its history would indeed cause the vampire considerable harm.

However, the modern vampire hunters do not understand how important this connection is and resort to technology to try and establish their superiority over the vampire's "child-brain" (Stoker 1996, 328), not realizing that the Count has equally strong natural powers at his disposal. The would-be slayers led by Van Helsing employ all manner of up-to-date communications technology, from trains to telegrams and from telephones to phonographs, to outwit and outmaneuver Dracula, but the vampire achieves the same and more through the nature of his being. As mentioned above, he manages fast and inconspicuous travel through bodily transformations—though luggage can be problematic—but personal communication is far more effective. As seen with Renfield, Lucy, and Mina, the vampire can directly and organically communicate with another's mind, sending and receiving messages over great distances. In fact, these messages are far more than the disjointed and financially expensive words transmitted by telegraph and, as seen with Mina's communications with the Count, can also include images and sensations. More importantly, these messages are inviolably secret unless someone chooses to share them and connect the sender and receiver in a way that even 21st-century technologies can only dream of.[16] This identifies one of the most vital parts of the vampire's organic communications—that they are personal and intimate—while those used by the hunters are extremely impersonal and often fragmentary—composed and transmitted by machines and recorded, written, typed, or stored in different places and different mediums. This is seen in the entirety of Dracula's affairs in Britain, where he grows to know people personally and intimately before sharing himself with them, and vice

---

16. As noted by my colleague Andrew M. Boylan, Van Helsing does manage to "hack" the Count's communication system but only by using hypnosis, though this is much nearer in nature to the "natural" sciences employed by the vampire than the modern technologies used by the vampire hunters.

versa, creating something of a holistic community that also tries to share his ecological values with others—Lucy as the "Bloofer Lady" is possibly the most obvious example of this, as she tries to recruit young children to the cause on Hampstead Heath. Arguably, this is the most dangerous aspect of the Count, as he is trying to deny the ravages that industrialization has brought to Britain and its society and return it to a more communal, social time, a past that the modernist project refuses to consider in its drive to the future.

Of interest within this is Dracula's use of natural resources. While many critics, such as Franco Moretti in *Signs Taken for Wonders* (1983), cite the Count as being representative of a kind of total consumerism and more in tune with the capitalist endeavor than the imperialists that defeat him, he is actually the model of a green economy. The means of communication mentioned above highlight this. The vampire hunters produce a huge amount of material, "a mass of typewriting," as mentioned by Jennifer Wicke (1992), and various photos, clippings, and even phonographic cylinders, while Dracula creates none. Apart from his initial communications with solicitor Peter Hawkins in Exeter, the vampire keeps his "footprint" as minimal as possible and either talks to people or communicates via his mind. Something of this is also reflected in his choice of means of travel to England. Where Harker used various steam trains and horse-drawn carriages, the Count chose a sailing ship that uses far less fuel and creates no harmful waste or emissions.

This equally applies to Dracula's dietary requirements, which his refusal to drink wine is part of. His hosts in England, and Harker when in Transylvania, require food and wine, along with the resources to produce it and the subsequent waste to dispose of. Dracula needs none of this, which is why he needs no working kitchen, electricity, gas, water, etc. or even the usual servants in any of his homes, resulting in an extremely small green or ecological impact on the environment around him. As such, his reliance on blood is the equivalent of utilizing a natural resource, which further helps to regulate the environmental impact of the human community on their surroundings. Even his methods of collecting or harvesting his food are shown as being extremely personal and humane, with Lucy and Mina almost being seduced to allow him to feed, with something similar happening to Harker back in Transylvania. In many ways this also reflects the differing natures of the opposing sides in *Dracula*, with the Count bringing a certain measure of holistic care and intimacy to his interactions with the local ecology, whereas the would-be slayers are less personal, coldly scientific, and more removed from the processes they undertake. Unsurprisingly then, Dracula's presence in England is extremely unsettling for the committed consumerists, who view expansionism and the subsequent exploitation of the environment as an ideological and technological right, and so they decide that he needs to be eliminated, not least because he has started to convert the women of their party[17]—the bearers of the empire's future—to his way of thinking. The slayers begin to destroy the Count's boxes of earth, breaking his connection to his memories, his homeland and its ecological strength, and so he must flee back to Transylvania to save himself.

The denouement of *Dracula* signifies a purposeful act of destroying not only the vampire's link to its ecosystem but also the imperialists' views on the environment and its resources in general. Dracula, as intimated earlier, is central to the balance of the environment around his castle, and he consequently represents not just the ideology of its ecological balance but the very land itself. Thus, when Van Helsing and the defenders of modernity attack the Count, they are also attacking the land and its resources. It is not surprising then that when Harker slashes at him and slices a large "wound" in his clothing, "a bundle of bank-notes and a stream of gold fell out" (Stoker 1996, 332). It is as though any colonial interference can only see a land, a landscape and an ecosystem in terms of money and what it is worth; the "blood" of Dracula as the "blood" of the land can only be viewed by capitalists as a monetary resource. The Count manages to escape Britain but is intercepted just before he can enter the safety of his home in Transylvania. Overpowered by his attackers, the vampire succumbs to multiple knife blows that cause his body to crumple into dust and vanish from sight (Stoker 1996, 409). The invaders see this as a moment of victory, but it signals Dracula's return to his home soil. As if to confirm this, the setting sun lights up the broken battlements of the Count's castle, and the wolves that are present withdraw to regroup. The land is once again complete, and the colonialists return home to an uneasy peace, one in which their faith in modernity is not as firm as it once was.

Nosferatu, F.W. Murnau, 1922 / *Nosferatu the Vampyre*, Werner Herzog, 1979

Murnau's film is arguably the first, if unofficial, adaptation of Stoker's novel—though Count Dracula did appear in the now lost Hungarian film *Drakula halála* in 1921—and, due to copyright issues, changes the names of the lead characters, the location, and even the date to be no longer centered on late-Victorian England but the fictional German town of Wisborg in 1838. As such, the tale is less about the conflict between a natural past and a

---

17. Mina is often identified as exemplifying a professional "New Woman" of the period, while Lucy is viewed as having far more "natural" views on marriage and female desire, seeing them as potentially open to more ecological, progressive views of the world around them.

technological future and more concerned with the mystery of the primal life-force of nature and humanity's incomprehension of it.

Wisborg is hardly a bustling metropolis and looks rural in comparison to late–19th-century London, though maybe not so different from Whitby. Murnau paints rural Germany as an almost dreamlike idyll with Hutter (the Harker character played by Gustav von Wangenheim) as its human representative. The only ripple in its calm—the worm in the apple, as it were—is Hutter's employer, Knock (Alexander Granach), a Renfield-like figure who seems to be in league with darker forces. It is Knock who arranges for Hutter to travel to Transylvania to deliver papers to Count Orlok (Max Schreck) to purchase a building in Wisborg that just happens to be directly across from where the young man lives; it is implied that Knock knows what the goal of this plan is. Hutter of course, being an innocent, happily agrees, though his wife Ellen (Greta Schröder) back in Germany almost immediately senses that things are not what they seem, and a shadow passes over her normally happy disposition. As the story unfolds it becomes clear that Ellen and Orlok are opposites who are connected in some way—life and death, light and shade configuring a form of balance within nature itself.

Hutter's journey to the vampire's lair is not as dramatic or as far afield as Harker's was, but it is still shown as a journey between worlds; if Wisborg shows nature being controlled and exploited by humans—there is much cultivated farmland around it—Transylvania is nature unleashed. Even as Hutter leaves Wisborg on horseback, he is seen entering a huge and mountainous panoramic landscape, a point of no return. This sense of going into the unknown, of a space over which he has no control and which might consume him, is strengthened when he stops at an inn before starting the last leg of his journey. The locals try to warn him of the dangers that await him, but, like Harker before him, Hutter takes no notice. However, as he spends his last evening in the comfort of the inn, it is intercut with scenes of the night landscape outside, of running horses and even hyenas—they were meant to be werewolves in the film—representing it as a wild, dreamlike environment totally alien to the world that Hutter knows. This impression is redoubled as he leaves the inn to rendezvous with the Count's coach in the mountains. Once there, the carriage arrives with a mysterious driver—Orlok himself—and departs at preternatural speed, with the film changing to a negative print so that the horses and landscape look like spectral, nightmarish versions of themselves; it is a world beyond this one where black is white and death is life. Once at Orlok's castle, things seem almost normal compared to the ride between worlds. The laws of the human world seem to no longer hold sway, or rather they are intensified to an extreme: The Count's world does not embody death; it is not the underworld but manifests the "dark" side of nature—at least when seen from a human perspective—and the desire, the necessity, to consume life to feed one's own. As such, it envisions an opposite world to that of capitalism where the laws of nature hold sway over the laws of mankind. Once inside Orlok's home, it transpires that Hutter's host already knows of the young man's wife, not through the picture he sees the young solicitor is carrying but through an organic, almost spiritual connection that joins such opposites—the yin and yang of life, as it were—innocence and light versus knowledge and darkness. Thus, the vampire prepares to leave his realm to go and meet her while leaving Hutter trapped in the world beyond our own. The departure of the Count causes ripples to course between the two worlds, affecting both Knock and Ellen—the two who are most intimately connected to the vampire. This also marks a reaffirmation that Orlok is not necessarily evil or death itself, just a part of the natural ecological order.

As the Count prepares to set sail, his boxes/coffins of earth are loaded onto the Vesta for transportation to Wisborg, and the dock is suddenly teaming with rats coming from the cargo; the vampire's connection to rats rather than bats is greatly emphasized here, and even more so in Herzog's later remake, not just as a literal translation of Nosferatu (as plague-carrier) but more as an expression of irrepressible and contagious life.[18] The scenes following this cut quite dramatically but reveal more of the links between the vampire and nature. Firstly, the film jumps to a text stating that "in those days, I should note Prof. Bulwer was explaining to his students the cruel habits of carnivorous plants"—the film uses the conceit that the tale, ostensibly of the year when plague came to Bremen in 1838, is being narrated by a historian called Johan Cavallius. Prof. Bulwer (John Gottowt) is then shown giving a lesson to his medical students, who are standing around a large box containing a Venus flytrap just about to catch its next victim. The shot then cuts to a closeup of the plant with an oblivious fly stumbling into its "jaws."

The text continues, saying: "They observed in horror nature's mysterious ways." Prof. Bulwer declares, "Like a vampire—isn't it?" The professor's narrative is interesting, not least because he is the Van Helsing character in this tale, as it correlates the action of a predator capturing its prey as vampiric—though this is also in part due to the fact of the predator being of a lower order (a plant) than its prey (an insect)—and also makes the vampire part of the natural order; this insinuation is repeated more than once by the natural scientist.

The scene then cuts to Knock being admitted to the local asylum, as the imminent arrival of the Count, his master, has driven him mad, though he equally seems to act as part narrator, part augur of what is happening. We return slightly later to Knock, now safely ensconced in the asylum where he

---

18. It is worth noting that the correlation to rats in the film, among other features of the vampire, can also be read as explicit antisemitism. See Giesen 2019, 103–10.

The Venus flytrap capturing an unsuspecting fly. *Nosferatu,* **directed by F.W. Murnau (Prana Film, 1922).**

is repeatedly saying "Blood is life. Blood is life" but then immediately attacks one of his guards. The scene suddenly shifts back to Bulwer and his students, seemingly looking in the same specimen box as before—they appear to be observing both the specimens and the mental patient—but with something else inside it. The professor says, "And this one here ... a polyp with tentacles ... transparent ... almost incorporeal ... almost a phantom." We see a tiny creature ensnaring what looks like a water flea in its tentacles and slowly crushing it. This again emphasizes the idea of predators being a valuable part of the natural order, both in the abstract but also in the events unfolding in present-day Wisborg, consequently citing the "natural" qualities of Orlok himself: a vampire, incorporeal, a phantom. This gives the creature a mysterious quality while simultaneously constructing it as a natural part of the environment. The scene then cuts back to Knock in his cell, pointing to a spider in its web that is busily wrapping its prey in silk—another vampiric creature entrapping its prey. This scene is quickly replaced by more text explaining the approach of Orlok and showing Ellen waiting for the return of her love, Hutter, on the beach. It is an intriguing set of scenes further linking Orlok to natural predators such as the polyp and the spider and the way they entrap their prey and slowly devour them. It further suggests that Ellen is the hapless victim unaware of the approach of the vampire and her inevitable death. However, this is not the case, as she is aware of the approaching menace and in a sense, as shown by the beach scene, longs for it to arrive. Indeed, it reveals her as a knowing player in the natural cycle of life and death and that she is not so much a victim as a willing and waiting sacrifice.

As the Count arrives and disembarks at the docks, so too does the pestilence that begins to ravage the town. The rats and the contagion are an extension of the vampire, parts of his essence that carry his "organic" energy into the world of humans—an ecological power that embraces both life and death, light and dark, as part of the evolutionary process. This is equally seen in the darkness that exudes from his body. Like Dracula before him, Orlok is not a human that transforms but rather a mutable concentration of matter/energy that takes on certain shapes. As such, his shadow is as real as the vampire itself, though rather than consuming blood it consumes light; in this way, it is not an absence of light but an active negation of it.[19] Consequently, once Orlok takes residence in his derelict property across from Hutter's home, his shadow makes contact with Ellen, not unlike the way in which Dracula would visit Lucy and Mina. However, here it is not as preparation for remaking her in his image but so they can restore some balance between nature and humanity. The shadow enters Ellen's bedroom and grips her heart, yet it is not enough to consummate their union; this can only happen face to face.

Ellen might be pure, but she is not unaware of what is occurring and has already investigated how to deal with the vampire's arrival. Once Hutter escaped Orlok's castle and returned home, he brought a book with him that he told his wife not to read, but, as noted in the film's pieces of text, Ellen could not "resist the bizarre attraction" of it and is overwhelmingly drawn to it as though beyond her control. As she reads, her destiny becomes clear:

> At night this same Nosferatu doth clutch his victim and doth suck like hellish life-potion its blood. Take heede that his shadow not encumber thee like an incubus with gruesome dreams. Wherefore no other salvation is possible but that a maiden wholly without sin maketh the Vampyre forget the first crow of the cock. Would she give freely of her blood [Murnau 1922].

Ellen realizes that, left uncontrolled, Orlok would spread his biological contagion over the world, changing the ecosystem to one that would not be welcoming to humans. This makes the end almost inevitable yet strangely shocking. It is evening and Ellen opens the window to her room to act as a lure for the vampire, convincing Hutter that she is ill and needs him to find Professor Bulwer at once. Sensing the time for their consummation

---

19. The film's inter-titles say that the vampire's shadow gets its strength from the cursed earth of the creature's grave. This apparently gives the shadow the appearance of grave soil but also marks a direct link between the vampire and its homeland.

is at hand, Orlok leaves his residence and makes his way to Ellen's room. This time, the vampire is directly connected to his shadow, and it appears to have even more presence, drawing strength from its undead master, and literally fills the woman's bedroom when the Count enters it. She offers her throat to him, and the vampire crouches by the bed to ingest that which he has so long desired: the blood, or life-potion, of one who is pure of heart. Throughout the narrative Ellen and Orlok are constructed as opposites; she feels genuine pain seeing flowers cut from their stems, while the vampire stalks and consumes its unwary prey; she is caring and nurturing, and he is destructive and voracious; she is light, and he is a creature of the shadows. The vampire's dependence on darkness is made even more pronounced by Murnau than it was by Stoker, for while Dracula was only nominally affected by sunlight, for Orlok it is lethal. Consequently, much of his attraction to Ellen is due to her connection to light and purity so that she can provide him with the only access to daylight he can ever have.[20] This begins to explain why he is so distracted, so fixated on his prey, on slowly and inexorably consuming every last drop of her blood (unlike the bloodbaths that accompany most vampire films since Hammer's *Dracula* in 1958, Orlok does not spill or waste a drop) that he does not hear the cock crow until it is too late. Just as he has drained every piece of light from Ellen, the morning sun enters the bedroom, expelling every bit of shadow from Orlok's body, causing him to disappear into nothing. Balance is then restored, and the contagion and the rats vanish with the vampire. The yin and yang of life and death, humanity and nature, of which the vampire is an inherent part, have achieved balance once again—though it can also suggest that enough death and disruption has occurred to slow the encroachment of modernity into the wilds of Europe. *Nosferatu* finishes with order being barely restored, but the later adaptation by Herzog gives no such reassuring conclusion.

Nosferatu the Vampyre follows much of the plot laid down by Murnau, though by the 1970s the worries around copyright no longer applied, and the characters revert to the ones used by Stoker, so Orlok is Dracula, Hutter is Harker, Knock is Renfield, and Ellen is Lucy Harker—this last being the director's poetic license and not uncommon in many vampire narratives where Lucy and Mina often seem to change roles. The connection between the vampire and other natural predators is not as explicitly shown here, though Count Dracula's affinity for rats is even more emphatically shown. This is especially evident when the vampire arrives in Wismar (Wisborg) on the now-deserted ship carrying his coffins. As the nefarious cargo is unloaded, the docks are

---

20. Something of this is seen in *Shadow of the Vampire* (Merhige 2000), an imaginative take on the making of Murnau's *Nosferatu* in which Orlok, who is a real vampire in the film, longingly watches a sunrise on a projection screen—the only time he has ever seen one.

virtually awash with vermin, a white flood of rats that literally washes over the town, infecting the population with the plague. In this sense, Dracula, while still configured as part of the natural order, is more symbolic of the unstoppable powers of death and chaos, almost an ecological bioweapon used to call a halt on humanity's march to an all-consuming, consumerist future; the setting of the film is now contemporaneous to Stoker's tale, and Wismar is a large, bustling mercantile town showing a world on the cusp of the 20th century.

Even more so than in the earlier film, Dracula's castle is shown to be the center of a domain that is not just in "the land beyond the forest" but almost not of this world. This is mainly seen once Harker leaves the inn to embark on the final part of his journey where, unlike the earlier versions, there is no horse or carriage to take him to the Borgo Pass, and he chooses to undertake the journey on foot. Kevin Jackson notes that Herzog "expands Harker's brief walk to Castle Dracula into a protracted and visually rich journey on foot, swathed in the prelude to Wagner's *Das Rheingold*" (Jackson 2017, 112); he further correlates this to the idea of the "Fussreise," or journey on foot, which is very much part of the German tradition of experiencing nature by traveling through it—an equal sense of the journey as life experience and going back to, or being part of, nature. Harker accomplishes both in this vastly extended journey through the landscape up and through mountains, by gurgling streams and waterfalls and watching the clouds pass overhead as day turns into night. It is very much in the German Romantic tradition and often feels like the paintings of Casper David Friedrich brought to life.[21] Friedrich's paintings and their recreation here are used not only to bring out the grandeur and power of the landscape as a terrain beyond the control and influence of humanity but also to endow the landscape with spiritual significance and evoke the sublime. The sublime here speaks equally of beauty and death, of a force that is destructive but also empowering or energizing (Burke 2015, 338). In many ways it also correlates to the abyss discussed by Andrew Smith, an overpowering blankness that has no regard for humanity while also hinting at the transcendental nature of the vampire. Dracula then is not just a predator, or even death, but part of the transformative nature of the environment itself, where destruction is part of creation and part of the biological energy (desire) that drives the ecosystem itself.

Unlike Stoker's book and Murnau's adaptation, Harker does not wait for the Count's carriage at a rendezvous point but continues walking alongside a river by the path, seemingly still lost in his thoughts. Suddenly, an ornate carriage draws up alongside him, not driven by the vampire in disguise but

---

21. Friedrich (1774–1840) was an extremely influential artist from the German Romantic period.

by an ordinary coachman who collects his passenger and delivers him to the castle. This part of the journey seems strangely ordinary, with no mysterious events along the way, emphasizing that the supernatural element occurred in the mountains and that we are already beyond the world of normal human experience. Curiously, Dracula is not yet aware of Harker's wife, and his interest only becomes aroused upon seeing the portrait that the young man carries with him. Consequently, the relationship between Dracula and Lucy develops along different lines, no longer a balancing of opposites, and its true nature is only revealed at the film's denouement. In this adaptation, Lucy realizes that it is Dracula who is causing the infestation of rats and its associated plague and that she is the only one who can save the townspeople. Again, unlike Murnau's version, Dracula seems to desire Lucy herself rather than what she represents—even though her present name Lucy, derived from Latin, actually means "light"—and so the story becomes more about humanity (human "beauty") being able to control nature (the environmental sublime). As in the earlier adaptation, she lures the vampire to her room, knowing that her own beauty and purity will distract him long enough from the impending sunrise. Thus, she pulls Dracula to her repeatedly until the sun rises and takes the vampire unawares. As the morning light enters the room, Dracula convulses and spasms, falls to the floor and slowly stops moving. Professor Van Helsing (originally Bulwer) runs into the room and stakes the prone figure to ensure that he is dead. Lucy, who is lying unmoving on the bed, smiles as she loses consciousness and dies.

Interestingly, the vampire does not vanish or disintegrate into ashes. He is not the incarnation of darkness or shadow—shadows play a substantially smaller part in this version—but more representative of the ecosystem trying to restore balance to itself. Lucy then becomes symbolic of humanity trying to coerce and seduce the forces of nature to its will. The smile on her face is one of self-satisfaction that once again humanity will have its way with the resources of nature. However, things are not as clear-cut as Lucy had thought, and her husband, who has been ill and almost comatose since returning from the Count's castle, is actually infected, and as the couple die upstairs, he awakens and escapes to carry on the vampire's quest.

The Forsaken: Desert Vampires, J.S. Cordone, 2001 / *From Dusk Till Dawn 3: The Hangman's Daughter*, P.J. Pesce, 1999

The setting moves from the mountains to the desert in this next example of eco-vampirism, creating what would seem to be an extremely unsuitable environment for the vampire, who becomes both a natural component of its ecosystem but also a representation of its inherent dangers and antagonism to human interference and development. Brad Sykes notes that the relation between vampires and the desert is "inherently animalistic ... mythical connections to bats and wolves (both of which are Indigenous to the Southwest and hunt at night) ... nomadic beings, constantly on the move or on hiding" (Sykes 2018, 182). Indeed, the linkage in film between the two started relatively early with *Curse of the Undead* (Dein 1959) and the Mexican movie *El pueblo fantasma* (Crevenna 1965), which featured vampire cowboys, though *Billy the Kid Versus Dracula* (Beaudine 1966) is the first to include the Count himself. *Billy the Kid Versus Dracula* is a curious film, not least for its uses and abuses of vampire lore. Dracula, played by John Carradine, sees a picture of Betty Bentley (Melinda Casey) and instantly decides he must have her, but as it happens she is the fiancée of Billy the Kid (Chuck Courtney)—he may be an outlaw, but he's an American one and so naturally defends his country from outsiders. The vampire seems utterly untroubled by sunlight and the usual deterrents but has an extreme reaction to silver, which brings about his eventual downfall. His death provides the most interesting part of the film, as it sees him staked with a railway spike, signaling a victory of modernity (Page, Weiner and Miller 2012, 59) and technology and the exploitation of resources over nature and the environment. This begins to delineate the way in which the vampire, as a symbol of difference and otherness, is always associated with the alienating and alien nature of the desert, while the hunters, as Western heroes on the cusp between worlds, are representative of the nation and futurity, a schematic that continues almost unaltered into the 21st century.

The desert in the films discussed herein is particularly American and usually on or near the border with Mexico. This situates it as a place that is often American and yet other, not unlike Transylvania in its ambiguous Europeanism. The resulting effect complicates the positioning of the vampire in these films, seeing him at times as a naturalized American and at others as a manifestation of traumatic history breaking into the present. Similarly, as noted by Stacy Abbott, the American desert represents the subgenre of the "South Western" as uniquely "expressive of Hispanic cultural survival" (Abbott 2007, 167). The self/other dichotomy is qualified further by Sara Gwenllian Jones, quoted in Abbott, who comments that the desert vampires in *Near Dark* (Bigelow 1987) "seem to erupt not from the European past but from nature itself, from the unruly wilderness that lies just beneath the surface of the farmed and settled American Midwest" (Abbott 2007, 165). This is of particular importance as it sees the landscape as a source of wildness whose physical expression becomes manifest in the body of the vampire and which is seen in both films discussed here. In many ways, this embodiment configures the desert as uncontrollable in and of itself, regardless of where it is. Notwithstanding that the various hunters and slayers that enter it may themselves signify a particular nation, the environment exists beyond their control and classification.

The two films discussed herein showcase this conceptualization of the desert, as both privilege opening and establishing shots of the landscape as outside of human control, a force unto itself and often sublime. Each film also shows a different expression of the vampire as part of the ecosystem. *The Forsaken* is similar to films such as *Near Dark*, where the vampires take advantage of the large open spaces of the desert and the roads that spiderweb across it. *The Hangman's Daughter*, not unlike *Vampires* (Carpenter 1998), *From Dusk Till Dawn* (Rodriguez 1998), and *Dracula III: Legacy* (Lussier 2005),[22] somewhat reverses that approach by creating a lair or honeytrap that draws unsuspecting victims to the vampires.

The Forsaken features a young and hopeful Jonathan Harker–type character, Sean (Kerr Smith), who is tasked with driving a car from California to Florida with strict instructions that prohibit picking up hitchhikers, which he inevitably ignores. The establishing opening shots of the film see the young man as a city dweller, chaotic and at turns boring but in his element in the city. As he begins his journey away from the metropolis, the skyscrapers are replaced by sheer-sided canyons in the weathered and worn landscape that is a precursor to the wide and wild expanses of the Arizona desert. The roads here are vitally important, and at the beginning of the journey, they act almost like a train carriage in that they seem to carry the car through the landscape, providing panoramic views of the surroundings while keeping the "passenger" separate from them; Harker has similar experiences as he starts his journey to Transylvania. However, as the adventure continues, the roads become less sheltering and begin to control and direct where the unsuspecting traveler may go. As if from out of nowhere, one of the car tires gets a puncture, forcing Sean to pull into the nearest garage. Once there, he discovers his wallet is missing and must pay for repairs from a small amount of money he had put aside for his sister's wedding that he planned to attend once the car was safely delivered. Thus, when a young man, Nick (Brendan Fehr), offers to pay for his fuel if he gives him a ride, Sean begrudgingly agrees. Unsurprisingly, it is not long before things start going horribly wrong and they run into a group of desert vampires led by Kit (Johnathon Schaech), who is something of a Dracula figure, Pen (Simon Rex), who also looks after their car during the day and acts as the Renfield character, and Cym (Phina Oruche) and Teddy

---

22. This last one is unusual, as the "desert" is a hugely extended wasteland in a future, war-torn Romania and features Dracula's castle at its center. A desert scene also features at the start of *Blade: Trinity* (Goyer 2003), which is supposedly in Iraq and contains the final resting place of the "real" Dracula, who lures vampires from the present to retrieve him.

Chapter 1. Dracula the Environmentalist

(Alexus Thorpe), Kit's two vampire "brides." It further transpires that Nick is a vampire hunter who has been bitten but keeps himself from turning by self-administering a special drug and who is trying to hunt down Kit in the belief that he is the vampire that infected him—in this film, if one kills the "source" vampire, all those infected by them who have not yet fed on blood will have their humanity restored.

The vampires seem to appear as if from nowhere or as if they had always been there. As noted by Jones above, the vampires in *Near Dark* seem to "erupt ... from nature itself" and do not appear until Sean and Nick are in the desert proper. Here, there are no large towns, only seedy motels, roadside bars, and solitary houses. As in *Near Dark*, the vampires are always mobile and live primarily in their car but sometimes sleep in hotel rooms, barns, etc. and seem to be made up of a ragtag mix of people whom the leader, Kit, has picked up along the way: Cym appears to be African American, though could just as easily be European; Teddy is a white, Midwest American; and Pen seems to be a low-IQ redneck[23] who is still human and acts as the group's protector during the day. Consequently, they form something of a stereotypical, patriarchal, supremacist view of a group representative of America itself being led by the more "superior," intelligent, and white leader. Kit, for all intents and purposes, seems to be American but is given a rather convoluted backstory for the film that sees him as one of eight French knights who were saved by a demon at the battle of Antioch (1098) during the Great Crusade. This resulted in his becoming a vampire. Subsequently, the knights parted ways to different corners of the world, and one assumes Kit came to the Americas when the French colonized part of it during the 16th century; his anti–British roots represent the founding of modern America.[24] The vampire leader is hundreds of years old, and perhaps due to spending such a large amount of time in the desert, he seems to have a kind of symbiotic relationship with it. Something of this is apparent when Kit picks up a rattlesnake and purposely forces its fangs into his arm. The poison is shown to be coursing through his bloodstream in the pronounced vessels pulsing in his arms—it is almost as if he is given life by the desert itself, making him part of its ecosystem as he takes on the characteristics of a desert snake and the larger environment.

Indeed, the desert vampire's predatory nature brings to mind the scenes of polyps and flytraps from *Nosferatu* as they lure victims into their traps. As the vampires track their prey, they seem to materialize out of the darkness of the desert next to stray travelers on the roadside or larger groups that stray

---

23. A type not uncommon in horror films and who often fills the role of psycho-killer as much as aiding and abetting one.
24. There is much here that intersects with the notion of "South/Southern Gothic" and historical otherness (See Cothren 2015).

from the beaten path. In one instance, they encounter a group of adolescents drinking beer around a fire with no idea of the danger they are in; they antagonize their mysterious visitors until it is too late. Another time, two young men in a truck think they can prey on the young girl by the road (Cym) but discover too late that it is they who are the naïve victims. These examples play on the idea of humans who think the desert is a safe, friendly place that they can exploit to their advantage but realize too late that it is wild, savage, and untamable with no respect for human life other than as a food source. In this sense, the desert purposely allows for only a limited incursion of humans into its ecosystem and disposes of those who stray from those "allowed" paths.

The Forsaken also implies that the vampires make occasional hunting trips to small towns on the edge of the desert as a warning to those trying to expand into the desert terrain. Indeed, this is where Megan (Izabella Miko), an infected girl that Sean and Nick rescue, was attacked. A similar pattern emerges in *Near Dark* with the vampires virtually patrolling the edge of the wilderness to keep humanity at bay. Consequently, the vampires become almost correlated to the sands of the desert itself, pushing against the growth of civilization at its edges and swallowing those foolish enough to enter its domain. The vampires finally succumb to the hunters, but the ending is not complete, as another surviving knight is still alive and resisting the forces of human exploitation.

The idea of a domain consuming those foolish enough to enter it forms the basis of the next film, *The Hangman's Daughter*, the third cinematic installment of the *From Dusk Till Dawn* series. Each of the three films has a focus on the desert: *From Dusk Till Dawn* (Rodriguez 1996) begins as a heist film but ends up in a bar on the borderlands of the Mexican desert; in *From Dusk Till Dawn 2: Texas Blood Money* (Spiegel 1999), a gang of criminals rob a bank in Mexico, but one of their number has visited that self-same bar with catastrophic consequences for all; the final film in the series focuses on the same bar, The Titty Twister, but a hundred years in the past to reveal more about its true nature and its relation to the landscape around it. It tells of various groups of disparate strangers that find themselves drawn to a mysterious bar in the middle of the desert. Of note in the different parties is the real-life author Ambrose Bierce (Michael Parks), who is trying to locate the Mexican revolutionary Pancho Villa, and Esmeralda (Ara Celi), the eponymous hangman's daughter who is also the girlfriend of bandit Johnny Madrid (Marco Leonardi). Their respective journeys are accompanied by amazing panoramic views of the desert—a vast, undulating, desolate expanse that is often silhouetted against the blazing sun.

Bierce's carriage gets bogged down in sand, and therefore, he and the two other travelers accompanying him have to proceed on foot through the parched landscape. Johnny and Esmeralda on horseback escape assorted

Chapter 1. Dracula the Environmentalist

The panoramic desert surrounds Marco Leonardo as Johnny Madrid and Ara Celi as Esmeralda in *From Dusk Till Dawn 3: The Hangman's Daughter*. Directed by P.J. Pesce (Dimension Films, 1999).

bands of vigilantes and troops trying to catch them, including her father, and finally stumble across the mysterious inn. Johnny finds himself particularly confused, as he has been through the area many times but never came across the inn. This increases the sense of destiny, and indeed the desert itself, drawing in unwary victims as a focal point of the ecosystem's revenge on humanity, particularly those who are exploitative and self-interested, as the groups have variously shown themselves to be. Initially the bar is empty, seemingly deserted except for the sudden appearance of barman Razor Charlie—played by Danny Trejo, who appears in various guises throughout all the films and the subsequent television series—but as soon as the sun goes down the bar is suddenly extremely lively. On the same site as the original Titty Twister, this inn is an earlier incarnation that acts as something of a bordello drawing in men and the disreputable from all around the region. As soon as Esmeralda enters, though, she is accosted by the "owner," Lady Quintila (Sônia Braga), who proceeds to lick the girl's face and declare her as her daughter, which is proven to be the case as the story unfolds. After this revelation, however, a fight breaks out in the bar, and the blood spilt during it causes all the girls and bar staff to change into vampires and attack the human customers.

The mythos of the vampires here is different from that of the later television series in which they are related to an Aztec snake god. In the film, they are intimately related to the fauna of the desert, just as Dracula was to the creatures in the environs of his castle. Consequently, when the vampires morph into their vampire form, they are partly bestial and serpentine, particularly in their respective "vamp-faces." In various scenes we see them change into large bats to attack a group of humans escaping through the dungeons under the bar, and when one vampire gets his head cut off, a new, cobra-like one appears in its place. As with Stoker's Count, the vampires in this film are intimately connected to their surroundings. They might not actively push back against the human incursions at the edges of their domain, but, as part of a symbiotic relationship with the desert, they dispose of humans who are foolish enough to venture into its midst. As the story draws to a close, Esmeralda is revealed to be none other than Santanico Pandamonium—the star of the first film and television series—the next in a long line of matriarchs who have ruled over the bar and the race of desert vampires that owns it. Madrid and Bierce escape, leaving the girl behind, and ride off into the landscape. The camera pans away into the sky to give a now-familiar aerial view of the bar—both previous films end with this aerial shot—to reveal the building as the top of an ancient Aztec pyramid with its rear half open to the world. The temple emerges from the side of a large sand dune, and rather than being buried or hidden by it, it seems to literally grow out of the sand as though it is an embodiment of the desert itself—not unlike how Count Dracula's castle grows out of the Carpathian mountains—a natural growth or concentration of sand that manifests its abhorrence of the human invaders who have dared disrupt its ecosystem. This equally suggests that the temple has purposely provided a home for the vampires in a relationship that is older than the American or Hispanic civilizations that have attempted to exploit the land for their own ends.[25] Although the two humans appear to escape unscathed, the desert has the last word, and as the film ends it cuts to a scene of Bierce sitting in a present-day bar, telling his fantastic tale to an unbelieving listener.[26] To prove the truth of his story, Bierce transforms into a vampire and rips out the man's throat, intimating that the desert no longer waits for humans to come to it but more actively intervenes in the world of humans to protect itself.[27] Desert vampires are thus strangely suited to the inhuman environment of this desiccated, sun-scorched landscape, expressing both its constantly shifting nature as well as its permanence and unremitting savagery. The next films examine a diametrically opposite environment but one with very similar characteristics.

---

25. Something of this is also seen in the film *Priest* (Stewart 2011), set on an alternate world where vampires are shown as an indigenous, non-human species that lives in huge underground hives in the deserts (reservations). However, they rise up against their city-based, human oppressors. This rural/city dichotomy is very much seen in *The Forsaken*.

26. This provides a fictitious answer to the actual mystery surrounding Bierce's disappearance while accompanying Pancho Villa's army in 1913.

27. The television series *From Dusk Till Dawn* (Kurtzman, 2014–present) more directly links the vampires to the indigenous Mexican culture that has infiltrated all strata of the modern American world.

30 Days of Night, David Slade, 2007 / Frostbiten, Anders Banke, 2006

In many ways, vampires should be more at home in snow than they are in sand: *The Twilight Saga* shows they are cold-blooded and hard as stone in *The Twilight Saga: Eclipse* (Slade 2010), where Edward is perfectly at home in the snow and needs the help of his romantic rival Jacob to keep the human Bella from freezing to death, and Roman Polanski's *The Fearless Vampire Killers* (1967) sees a large vampire community happily traversing the snowy landscape of Eastern Europe. However, the migration of vampires to the urban environment sees them largely eschewing such extreme environments, and films like *Stake Land* (Mickle 2010), possibly following the logic of zombie apocalypse movies such as *World War Z* (Forster 2013), show that colder climates, such as areas of Canada, are not preferable for the undead and can offer some form of sanctuary for humans.[28] However, this has not stopped certain frozen and freezing landscapes from protecting themselves from human incursion through the manifestation of vampires and vampiric entities. Possibly one of the most dramatic of these, and explicitly exemplifying the horror of the white abyss, is *30 Days of Night*.

Not unlike *The Hangman's Daughter*, this film by Slade is actually part of a larger franchise originating from a series of three comics by Steve Niles and Ben Templesmith (August–October 2002) and including two online promotional mini-series (a seven-episode prequel titled *Blood Trails* [Garcia 2007] and a six-episode sequel titled *Dust to Dust* [Ketai 2008]) and a movie sequel titled *30 Days of Night: Dark Days* (Ketai 2010). *The Twilight Saga: Eclipse* will be looked at as a standalone film here, as it is in that form that its ecological focus is best teased out. *30 Days of Night* takes place in the Alaskan town of Barrow, which is the northernmost city in the United States and north of the Arctic Circle, which experiences a continuous period of 30 days of arctic night every year.[29] This setting provides the perfect opportunity for light-sensitive vampires to run amok in the town, causing death and mayhem. This idea of taking advantage of temporary or permanent cloud cover is not uncommon in vampire films and features in movies such as *Twilight* (Hardwicke 2008), where the vampires move to Forks, Washington, because it is the region with the most cloud cover during the year[30]; *The Matrix* (the Wachowski Brothers 1999); and *The Strain* (del Toro and Hogan 2014–17), in which vampiric entities use nuclear weapons to create dust clouds to cover the landscape from

---

28. This is rather disavowed in *Stake Land II: The Stakelander* (Berk and Olson 2016).
29. The town is now known as the City of Utqiaġvik and was originally founded by the Iñupiat, an indigenous Inuit community.
30. The area sees an average of 212 days of rain (measurable precipitation) a year.

sun. As Slade's film begins, it makes much of Barrow's isolation, as it consists of "80 miles of wilderness" and portrays a lone figure walking across a vast expanse of deep snow with what looks like a huge snowstorm approaching in the distance. The man arrives at the crest of a hill with Barrow below him covered in huge, ominous storm clouds to be replaced by an equally troubling text declaring "the last day of sun" (Slade 2007). Not unlike *The Hangman's Daughter*, there are many magnificent panoramic images of the snowy landscape that stretches away for as far as the eye can see; Barrow is a town at the mercy of the wilderness.

The impending period of darkness causes the town's inhabitants to become increasingly anxious and frantic. This conceptualizes it as both a regular occurrence—there is a certain familiarity in the traumatic separations of friends and families shown—and oddly catastrophic, as seen in the desperate rushing of the local sheriff's ex-wife, Stella (Melissa George), to catch the "last plane out" before night falls and the approaching storm arrives. This is further emphasized by her frantic drive to the airstrip and crash into the side of a snow plough, almost killing herself. As Peter Hutchings notes, "The town itself is shown as isolated and embattled, with polar night representing the climatic environmental assault on the community" (Hutchings 2014, 62). The sense of impending doom is further emphasized by a series of malicious occurrences perpetrated by the stranger seen at the start: all the mobile phones in the town are stolen and destroyed, the police helicopter is sabotaged, and all the sled dogs are killed. In fact, the approaching bad weather, the falling darkness, and the mischief caused by the mysterious stranger all become conflated as a product of the environment itself; it is rising up against the human intrusion represented by the inhabitants of the town.

As the snow starts to fall, the vampires begin to edge into the town, snatching and killing people, beginning with the communications office and the powerplant, which both constitute the focus of the human intrusion into the arctic environment. The powerplant also has an oil pipeline running through it, revealing it to be a particular drain of the ecosystem's resources, in addition to the negative effects of using/burning fossil fuels on the polar regions of the planet. Here, the vampires "disappear" their victims—a feature very specific to the film—by grabbing and pulling them away so quickly they appear to have suddenly vanished. This is not the slow encroachment of the desert or the eroding of borders but the cracking glacier, the suddenly opening crevasse, and the snapping maul of the abyss that swallows a person whole. This posits a dramatic shift to the vampires manifesting something darker than the usual undead, as noted by M. Jess Peacock in relation to one of the townspeople who appeals to God but to no avail, casting the vampires as creatures of "greater mystery and ferocity, unable to be contained by the traditional sacred icons" (Peacock 2015, 82). In this sense, the vampires

Chapter 1. Dracula the Environmentalist

Vampiric embodiment of the arctic ecosystem coldly portrayed by Andrew Stehlin as Arvin in *30 Days of Night*. Directed by David Slade (Culver City: Columbia Pictures, 2007).

embody the destructive power of the arctic ecosystem—not just a never-ending whiteness that stares back but one that consumes you instantly.

This idea is reinforced by the look of the vampires themselves, who methodically begin to hunt and kill the inhabitants of the town. They are deathly white with jet black eyes and rows of sharp teeth in their mouths, and their speech is a series of guttural clicks and squawks. Unlike the desert vampires, they do not seem to mimic or borrow attributes from any of the local fauna but instead represent an inhuman embodiment of the environment itself: cold, vicious, and voracious.[31]

They kill most of the townspeople except for a small group of survivors led by the sheriff, Eben Oleson (Josh Hartnett). The sheriff manages to get some of the survivors to safety, but as the first sunrise approaches, he realizes that the vampires are planning on setting fire to the town to cover up what they have done. In a last-ditch attempt to save the remaining survivors, including his ex-wife, Eben challenges the leader of the vampires, Marlow (Danny Huston), to a fight, knowing that victory would save everyone. To

---

31. Something like this is seen in *The Colony* (Renfroe 2013), where a future Earth is consumed by another ice age. The few human survivors live underground and have to combat not only the cold but also the gangs of vampire-like savages that eat flesh. Like the *30 Days of Night* vampires, they are animal-like and feral and possess rows of sharp teeth, bringing the violence of the ecosystem on the surface to the chambers underground.

accomplish this, Eben injects himself with infected blood so that he will have the increased power and strength of a newly turned vampire but times the combat so that he will be partially human and will therefore still have control over his actions. In part, this sees the sheriff taking on the strength of the environment around him so that it will cancel itself out.[32] Inevitably, Eben defeats Marlow, dissipating the forces of destruction that were unleashed by the ecosystem, but the level of destruction caused is sufficient to stem any flow of outside human exploitation, particularly seen in the destruction of the power station. Eben knows that ensuring the continued safety of the town and his loved ones requires his own demise. He thus persuades Stella to take him to the edge of town to watch the sunrise, which sees him, like Orlok before him, disintegrate into ash (if a little more painfully). This is something of a final sign of balance being restored with the returning sunshine, not unlike that in *Nosferatu*, suggesting a new time is beginning, though most of the vampires survived and are able to deliver their ecological justice at any point in the future.

The narrative of *30 Days of Night* lacks history insofar as it is very much of the moment and without precedent. The environment in this film has decided to act in a swift and catastrophic manner to limit further incursion, or to at least make the human invaders think twice about continuing their current course of action and lessen their sense of superiority/security within the natural order.

The next film to be looked at shifts this perspective, seeing eco-interventions as part of an ongoing historical trauma in which the environment disrupts human exploitation of its natural resources. This idea goes back as far as *The Thing from Another World* (Nyby 1951), in which members of the U.S. Air Force find a buried craft under the arctic ice and return with what seems to be its only crew member to conduct tests on the craft and the crew member in their base. The crew member turns out to be an unknown (alien) plant-based lifeform that feeds on blood and destroys most of the base before being killed. The 1982 remake by John Carpenter increases the historical aspect of the craft and makes the "plant" far more destructive and potentially catastrophic for the human race, thereby highlighting its role as an organic apocalypse to reset the Earth's ecosystem. *Frostbiten* uses many elements of this "traumatic return" but with less cataclysmic intentions for the human race.

---

32. This is probably one of the most contentious parts of the film in terms of the actor chosen to play Eben. In the comics, the sheriff is called Eben Olumein and is seen to be of indigenous descent, and this would work much better in terms of harnessing the power of the ecosystem around the town. Unfortunately, the film uses American actor Josh Hartnett, changing his name to Eben Oleson and losing the familial links to the area, depicting him as more representative of the outsider forces exploiting the environment rather than its protector.

Frostbiten begins in 1944 in a very snowy Ukraine during World War II, where a division of Swedish volunteers to the Nazi cause, 5th SS Panzer Division Wiking, are running away from the advancing Red Army.[33] They stumble upon a seemingly abandoned cabin and decide to spend the night there away from the freezing conditions outside. Once inside, the surviving soldiers begin to ponder the fate of the cabin's original inhabitants and realize that the snow piled up all around it over the doors and windows means that they could not have left and must still be in there. As the soldiers begin searching, they discover a crypt under the cabin and the undead vampires that live there, resulting in the soldiers' untimely demise but for one survivor. The survivor, who is infected, kills his remaining companions, takes a box from the cellar and trudges off through the snow with it. There is something of the Titty Twister here, with the cabin being both a focus point of the surrounding ecosystem and a trap for the unwary, and as with Bierce, it sends feelers out into the world to influence the wider environment.

The scene cuts to present-day Sweden and shows doctor Annita (Petra Nielsen) and her daughter Saga (Grete Hovnesköld) moving away from "civilization" in Stockholm to a town in Lapland, Northern Sweden, just as the polar night is about to begin—not unlike in *30 Days of Night*. Annita has chosen to move because her idol, renowned but reclusive geneticist Professor Gerhard Beckert (Carl-Åke Eriksson), works in the hospital there. The town is barely shown other than its inhabitants, who seem rather bored of living in a place so far from excitement, a gripe that they soon regret. Unlike in Barrow, there is no panic or sense of impending doom as the sun vanishes for a month. The town appears "unaffected" by the oncoming winter night (Hutchings 2014, 62), but Outi Hakola views it slightly differently, seeing the town as "threatened by uncontrollable death" (Hakola 2015, 206), and while this is described more in terms of social positioning, it sees the community ripe for ecological disaster. The only oddity is the death of a young boy on the first evening of darkness—strangely reminiscent of one of the kill scenes from the later Swedish film *Let the Right One In* (Alfredson 2008)—which is a sign of things to come. Annita begins work at the hospital and soon meets Professor Beckert and his only patient, a girl who has been in a coma since her car crashed over a year ago. Indeed, the patient becomes something of a metaphor for the environment itself, being artificially kept in a state of unconscious stasis while unseen nefarious experiments and tests are going on. This point is brought home later when the coma patient awakens momentarily and bites Annita. This forces Beckert to knock her out and tie her up, as he knows Annita will soon become a vampire, just like the girl that bit her. It transpires that Beckert is the soldier who

---

33. The Division actually existed and was made up of volunteers from Scandinavia and Northern Europe.

escaped from the cabin all those years before and who took a vampire child away with him in the box he transported. Supposedly, he began his career as a geneticist to find a cure for the condition but realized after some time that this was a hopeless endeavor and so devoted his more recent research to evolving the vampire into something new that could take over the world. Indeed, the recent murder happened because of his own "evolution" into this new species of vampire—unlike the vampires in *30 Days of Night*, Beckert can transform himself to a certain degree; like Stoker's Dracula, he can change his human appearance from young to old and even transform it entirely, as well as turning into his true vampire form. This more clearly shows the excessive animalistic/environmental energies he embodies as he becomes a huge, bat-like creature, minus the wings (though the earlier scene of young boys being lifted into the air suggests he might be able to fly or is at least extremely agile)—very similar to the Subsiders in *Daybreakers* (the Spierig brothers 2009). He develops huge, claw-like hands, loses all body hair and grows large, upwardly pointing ears, glowing eyes, and rows of large teeth. When Annita wakes up, he tells her how he experimented on the coma patient in order to perfect a drug in the form of glowing red pills that will turn humans into these new vampires. The professor invites Annita to join him, pointing out that she has only managed to overpower him because she is changing into a vampire. She refuses, but he tells her that she is settling for less:

> BECKERT: I tried to cure her. But she was too stubborn [the coma patient]. I'm the latest model. She's stuck in the past. I'm close to perfection. She's just a stepping-stone for the new species. Just like you. You're stuck in the past too [Banke 2006].

The statement is worth looking at more closely, as it differentiates the kind of ecological impact of the two kinds of vampire, the reactive and the proactive. The girl in a coma and Annita have the original infection brought from the frozen forest in Ukraine—the source of that infection is still alive and locked away by the professor—which envisions the vampire as a limiting factor to human incursion, a trap for the unwary and those in places they should not be. Beckert is far more proactive in his interventions, pushing back against humanity to an inevitably apocalyptic extent. In the ensuing struggle, Annita manages to kill Beckert; she stakes him, which has little effect, but as she removes it and he is bathed in light from an ambulance, he disintegrates into dust.[34]

Meanwhile, Saga was attending a party where the red pills were distributed among the guests and suddenly finds herself surrounded by vampires,

---

34. It is not totally clear in the film if it is the combination of the vehicle lights and the stake or just the stake being removed that finally destroys the vampire.

which are of the proactive variety and which seem unable to control their natural instincts. Somehow, she manages to escape the vampires and is saved by her mother, who is driving an ambulance. As Saga begins to take in her surroundings, she notices a little blonde girl sitting next to her. After a moment, the little girl says, "we're sisters now and mummy is going to look after us." Saga looks over her shoulder into the front of the ambulance and sees her mother's face in the rear-view mirror. Her eyes are glowing red in the darkness. As they drive away from the environmental disaster of the town, the inference is that the trauma from the past (the reactive vampires) will remain just that: an occasional recurrence that will keep the ecological balance within its vicinity (limit the negative effects of humanity), a blast that will take the world (immediate environment) back to its past.[35] Part of the devastating effect of the past is that the present is totally unprepared for it.[36]

The vampires here tend to embody a snowy environment rather than being natural residents of it, though as mentioned, some do take on certain characteristics of the fauna of such landscapes. In contrast, the next examples feature vampires that live in the habitat they represent, making them more evolutionary in nature and as wild as the ecosystem that supports them. Accordingly, the approach of such ecosystems is not to obliterate humanity, as in *30 Days*, but to assimilate it, not so much that humanity becomes one with it but that it becomes something new.

The Black Water Vampire, Evan Tramel, 2014 / *Annihilation*, Alex Garland, 2018

Tramel's film is heavily influenced by *The Blair Witch Project* (Myrick and Sanchez 1999), the breakout movie for handheld camera and faux-documentary filmmaking. Like *Blair Witch*, *The Black Water Vampire* shows

---

35. Something like this is seen in the slightly later Swedish film *Let the Right One In*. Here, a seemingly young girl, a vampire who is created by the snow, brings disruption to a Stockholm suburb to remind it that that it can never escape the trauma of the past.

36. Something of this is seen in the earlier Finnish film *Valkoinen Pura* [The White Reindeer] (Blomberg 1952), which shows a remote Lapp community in the far north of the country where a local woman unknowingly makes a blood-pact with the environment, and only traditional methods can dispel the vampire and restore balance. This idea is taken up in the film *Blutgletschern* [Blood Glacier; also known as The Station, 2013] by Marvin Kren in which scientists are exploring in the Austrian Alps and discover a mysterious red liquid coming out of a glacier. Even more so than *The Thing*, it plays on ideas of humanity imposing itself in places it should not be, as this ancient fluid begins to affect the local wildlife, turning them into monstrous, prehistoric versions of themselves, all of whom seem intent on killing the humans in their vicinity. The anxieties produced in the human population are maybe best summed up in the tagline "winter is coming" from the series *Game of Thrones* (Benioff and Weiss 2011–19), where the White Walkers, undead revenants from the past, threaten the whole of civilization by returning it to a frozen wasteland to curtail the ambition and greed of civilization.

the local woodlands to be far more hostile to humanity than one would think, though *Blair Witch* aims for more supernatural reasons for the catastrophic/surreal events at its close. In contrast, *The Black Water Vampire*, while featuring what are often thought of as supernatural beings as its theme, depicts these beings in a far more evolutionary incarnation. The backstory to the movie is that every 10 years, at least for the last 40, the body of a woman drained of blood is found in the woods near Black Water creek by the town of Fawnskin. A local is blamed and convicted of the murders and is currently in jail on death row. However, documentary filmmaker Danielle Mason (Danielle Lozaeu) does not believe he is guilty and enlists two friends, Andrea (Andrea Monier) and Rob (Robin Steffen), and a hired cameraman, Anthony (Anthony Fanelli), to join her investigation of the woods where one of the bodies was discovered. As they begin their investigations, they learn of a legend of vampires in the woods, which gives the nearby main road, Bloodsucker Highway, its name. On their way there at night, they swerve off the road when a mysterious figure darts out in front of them and disappears. They make their way to the town of Fawnskin, but the residents they meet and the landlady of the property they have rented act strangely in their presence. Indeed, as the story proceeds, it presents the kind of vampiric ecosystem seen earlier but transposed to pastoral New England and the primal forests hidden at its core.

Rather than reaching out to the world, the vampiric forces of nature protect the ecological heart that beats within the large areas of woodlands that are encroached upon by the students from the "big city," suggesting the same kind of rural/urban antagonism mentioned earlier in relation to *Priest*. The landscape is not the same kind of Romantic sublime seen around Count Orlok's castle, though it does create a very similar hallucinatory space, which is almost immediately experienced by the film crew as they enter it. The further they venture into the trees, the more lost they become, seemingly circling around on themselves, not unlike *Blair Witch*, so that it seems as though the woodlands themselves shift and move around them, slowly wearing them down and pulling them apart. When they finally reach the spot where the last victim was found, they realize that Rob has vanished, and although they spend much time looking for him, he seems to have been literally consumed by the woods. By this time, they are in snow-covered parts of the woodlands, with the sense of otherworldliness increasing the deeper they go. That night, they are attacked by vampires—large, bat-like creatures (very much like the Subsiders from *Daybreakers* or the vampires from *The Hangman's Daughter*) who drag Danielle off into the trees. During the group's search for her, they see one of the creatures hanging upside down in a tree and happen upon her standing naked in the snow once daylight breaks. She complains of something stirring inside her, and indeed there is a large lump in her abdomen, suggesting she might have been impregnated somehow. Anthony decides it is

Chapter 1. Dracula the Environmentalist

Brandon deSpain as the titular character in *The Black Water Vampire*. Directed by Evan Tramel (Los Angeles: Image Entertainment, 2014).

time for him to leave, and as he turns to walk away, he comes face to face with a vampire that kills him.

The two girls run but find themselves hemmed in and steered by various townspeople who seem to be forcing them toward the house of Raymond Banks, the man convicted of murder. Once they arrive at the house, Andrea is killed and Danielle is grabbed by some of the townspeople and held down to the ground—thereby increasing her contact to the environment—as the creature inside her bursts out. The scene fades to black to be replaced by a birthday party a year later. The room is filled with townspeople, and the camera is directly over a large iced cake with "Happy Birthday" written on it. The lens moves around the room bustling with happy, chatting people and finally alights on Danielle, who sits staring blankly into space like a mannequin. Next to her is a crib containing a baby vampire, which suddenly squirms, revealing a mouth full of razor-sharp teeth. The scene is purposely reminiscent of the ending of Roman Polanski's *Rosemary's Baby* (1968), which sees a group of New York Satanists arranging for the Devil to have a baby using an unsuspecting female victim but which is also very much about the environment of the city and the dark powers behind the urban landscape. This same feeling is transposed to the woodlands of Black Water, representing the inhabitants of the nearby town as intimately connected (making a pact) with the ecosystem so that the prosperity of one is ensured by the other to naturally repel outsiders and their unwanted influence, using any instances of incursion to further consolidate their pact. In this way, it departs from the scenario of *Blair Witch* where the environment defends itself—the witch needs no help from outsiders and personally takes care of intruders. In contrast, *Black Water* requires the active involvement of the human townspeople to ensure the survival of the vampires—almost seeing them as an endangered species—and the woodlands they live in. In this sense, the film is not unlike *Jug Face* (Kinkle 2013), which similarly sees a community founded around, and highly protective of, a vampiric entity—in this case a hole in the ground[37]—and sets itself up in opposition as the rural versus the urban environment.

Interestingly, the vampires seem to be the only creatures in the woods and do not seem to have a natural place to live in—unlike bats, they do not seem to live in a cave, although they do hang from trees. That said, the impregnation of Danielle and the slow aging of the newborn seem to suggest not only a biological foundation but a human equivalency in terms of aging. The legend of vampires in the area also hints that they are evolutionary creatures, making them an intimate part of the ecosystem rather than symbolic of it, as seen in many previous examples here. This dependence on the human community around them—making the entire town symbolic of the "Renfield" character—redirects the Black Water vampires away from the kind of "honeytrap" motif seen in some earlier examples in which they would kill or change all humans in their vicinity to instead use humans to increase or maintain their own numbers. There is a certain sense of assimilation involved here, though Danielle's comatose state speaks more of regression than any positive form of mental or physical hybridity. This kind of evolutionary hybridity, however, does form the basis of the next example, which is extremely aggressive in its assimilatory abilities.

While *Annihilation* cites the catalyst for the ensuing events as coming from outer space, it is possible to view them as an expression of the environment itself, symbolic of the ecosystem and its attempts to deter human incursion. *Annihilation* begins with the former but ultimately infers the latter as the mutation at its core spreads ever further. Also of importance in placing Garland's film here is the connection to the environment around Dracula's castle, as discussed at the start of this chapter, where the vampire's lair is situated at the center of an ecosystem that is separated from that outside of it. For Harker in Stoker's story, the surreal journey in the Count's carriage begins at the Borgo Pass, and for the same character in Herzog's film, it starts as soon as the "wanderer" enters the sublime landscape; both construct the environment as excessive: more fecund, more rugged, more spectacular. Even the animals seem otherworldly and excessive, as demonstrated by Murnau's hyena/werewolves and, in the official film of Stoker's novel by Tod Browning, armadillos, opossums, and cockroaches that appear to live in their own minia-

---

37. This is not unlike "The Transfer" by Algernon Blackwood (1911), where a patch of land seems to draw the life-force out of all who come near it.

ture coffins, suggesting a special and environmental hybridity that conforms with the "weirdness" of *Annihilation*. Indeed, rather in keeping with the idea of Dracula's castle as an "eye" at the center of its environment, the excessive ecosystem in Garland's movie has a lighthouse, a point from which the jouissance radiates and where the "vampire" lives. Emphasizing these similarities further, the "Shimmer," as it is called in the story, forms something of a dome over the area overseen by the lighthouse, and just as with Stoker, Murnau, and Herzog, the area immediately inside this is almost hallucinatory in the effects it has on humans entering it; as mentioned earlier, Murnau's film sees the image sped up and changed to its negative in an unsettling inversion of filmic reality. In fact, Murnau's film is also relevant in its use of "vampiric" microbes, the tentacled polyp, to emphasize how Orlok is part of the natural order. This maps onto *Annihilation* quite neatly, as it sees the very structure of organic life restructured in a way that seems to feed off other forms as its DNA is warped, twisted and invigorated in ways that exceed current human conceptions of what constitutes life on Earth.

Unlike with Count Dracula and Orlok, where the otherworldly region around the castle is prepared to wait idly and stay where it is while its sole representative is sent out into the world on its behalf, the Shimmer, as the new, supercharged bio-environment is called, slowly and inexorably moves ever outward, changing the world around it. Similarly to Stoker's tale, representatives of the outside world are sent into the "land beyond the forest" and, rather than bringing the matter to a successful conclusion, facilitate the incursion of the vampire into the wider world. The imperial travelers consist of a team of women, all scientific or medical personnel, chosen by the secretive Southern Reach organization to enter the alien realm and discover more about what it is and its possible purpose—they are the latest in the line of people sent in who were never heard from again but for one man from the last expedition named Kane (Oscar Isaac), who is slowly dying in a medical facility. To find a cure and save her husband's life, his wife Lena (Natalie Portman), a biologist, volunteers for the next mission into the Shimmer and joins four other women—Josie (Teresa Thompson), a physicist, Casie (Tuva Novotny), an anthropologist, Anya (Gina Rodriguez), a paramedic, and leader Dr. Ventress (Jennifer Jason Leigh). Their entry into the strange alien world is as mysterious as Hutter's carriage ride. One minute they are on the edge of the dome and the next—which is actually three days later—they wake up surrounded by amazingly luxuriant and exotic plants; if Nosferatu's realm is sublime because of its rugged, mountainous beauty that towers around Hutter/Harker at every turn, the Shimmer achieves the same effect with a green tidal wave of plant life that, metaphorically, washes over the expedition party. The Shimmer, as Elizabeth Parker observes, is "an enclosed, but ever-expanding and dome-like space in which all living things mutate and transmogrify at unprecedented rates" (Parker 2019, 95).

Unsurprisingly, none of the expedition's communications or locational equipment function, and they are therefore unable to contact the outside world or easily pinpoint their location within the Shimmer. They decide to head to the nearby coastline and work their way toward the lighthouse, which has been identified as the ground-zero point of the organic explosion, but before they have made substantial progress they are attacked by a large, aggressive gator-shark: an alligator with the teeth of a shark. Fortunately, no one is killed by it, but it is a curious hybrid, as the two predators do not share the same habitat. In many ways it seems a mix more predicated on human anxiety in the same way monster films such as *Piranaconda* (Wynorski 2012), *Sharktopus* (O'Brien 2010), and *Dinoshark* (O'Neill 2010) interbreed predators that are part of the current popular imagination as a way to excite audience numbers and play on cultural anxieties over biological science and an ecosystem out of control. As such, the hybrids, rather than making more efficient predators, are more directly aimed at unsettling or vampirizing the human population by feeding off their heightened emotional states.

Something of this is confirmed by the next hybrids they encounter once they make camp at the former base camp of Southern Reach, which has now been consumed by the ever-widening Shimmer. Once there, they discover tumor-like organic growths around the base and a video made by the previous expedition of one of their number being sliced open by Lena's husband, revealing the man's intestines squirming inside him like a huge snake as though they have a life of their own. Not long after watching this video the group are attacked by a bear-like creature that takes Cassie away with it and kills her. While the party are searching for the missing girl, they come across some peculiar trees that have grown in the shape of human figures and which they think is due to the random mixing of DNA within the Shimmer, though it can equally be read as the "vampire" at the center of the ecosystem experimenting with ways of copying and/or creating humans from its own hybrid organic matter—in a sense not unlike the plant-based creature in Carpenter's *The Thing from Another World*. After finding their companion's body, the scientists return to base, only for the "scream-bear" to return—named such as it is the size and shape of a bear but makes noises like a human screaming. It seems to have a human skull as half its face, appears to take on characteristics of its victims, and, when it returns to the camp, makes a noise exactly like Cassie screaming—an idea that was similarly used in *The Hunger Games* (Ross 2012), where various creatures could mimic human voices and songs. The scream-bear, more than some of the other mutations, is "an unforgettable demonstration of ecophobic dread, as the illusion that humans and nature can remain wholly distinct is harshly contradicted" (Parker 2019, 98). This more than the other mutations seems specifically directed to unsettle humans and, at least in terms of what is seen within the film, intimates that humans are their main/preferred prey or that the environment itself is using them to target human intruders, which is a view that gains much credence at the end of the story. The scream-bear manages to kill Anya, and the following day, Dr. Ventress decides to head out by herself while she still can. Josie discovers leaves and roots growing out of her body and joins the human-shaped plant forms, leaving Lena to make the final trek to the lighthouse alone, knowing that her body has already been irreparably altered.

This is particularly interesting in terms of the correlations between the ecosystem of the Shimmer and that around Dracula/Orlok's castle, as Harker is equally, and irreparably, transformed by the time he spends there. In many of the later film adaptations, particularly those where Renfield replaces Harker as the one who visits the castle (for example, Tod Browning's 1931 film), the hysterical zoophage barely regains any sense of who he was before entering the ecosystem, and death is his only escape. The area around the lighthouse is full of crystal trees surrounded by skeletons on the sand, although the building itself is covered in plant life. Lena enters it to find a scorched figure sitting against a wall in front of a video camera on a stand. The recording shows her husband holding a phosphorus grenade while sitting where the scorched figure is, telling someone off camera that he no longer knows who he is and asking them to look after his wife. The figure comes into view, and it is her husband, or a copy of him. Then the grenade goes off. Lena realizes that the man she met before entering the Shimmer was not her husband but a clone produced by the environment itself. There is a large hole in the floor of the lighthouse that Lena enters, where she finds the source/focus of what is happening in the ecosystem; she also finds Dr. Ventress there. The room/cave is of particular note as it is extremely womb-like, both a source of life and a consuming monster (vagina dentata)—certainly in terms of nature/ecosystem being signified as female and technology/consumerism as male. It resembles in appearance the birthing chamber of the xenomorphs from the *Alien* series of films, which further posits it as an organic space—the xenomorphs seem to make it from their own bodily secretions—but also one that is extremely aggressive toward humanity. Some of this is borne out but in less extreme ways than seen in the *Alien* films.[38]

Ventress has begun to mutate and gives herself to the source of the excess, disintegrating into a nebulous form that manages to absorb a drop of Lena's blood that has come out of her face. The organic source creates a blank copy of Lena from her drop of blood, a formless mannequin that appears to be made of mercury, reflecting and absorbing its human double. This process is not uncommon in vampiric entities, from the organic creature in *The Thing*

---

38. The environments built by the xenomorphs are most graphically seen in *Aliens* (Cameron 1986).

Lena's vampiric mirror. Natalie Portman as Lena (left) and Kristen McGarrity as Lena Double in *Annihilation*. Directed by Alex Garland (Hollywood: Paramount Pictures, 2018).

(Carpenter 1982), which absorbs the blood of any lifeform it is going to take the form of, to *Terminator 3: Rise of the Machines* (Mostow 2003), where the liquid metal robot from the future "tastes" its victims before mimicking them.

The organic doppelgänger mirrors the movements of Lena, slowly taking on her form up until the point that the pair hold onto the same phosphorus grenade, which then explodes. One of them runs out as the other begins to glow and becomes increasing formless before bursting into flames. The fire seems to trigger a chain reaction, and the crystal trees, the growth around the lighthouse, and even the Shimmer itself disintegrate into nothing. Lena returns to mission base to be debriefed on the events that happened in the Shimmer before it vanished. She suggests that it wanted to create something new, but she doesn't know what. Her husband's health has suddenly improved, and Lena is reunited with him. She asks him if he is her husband, and he replies that he does not think so. He asks her if she is Lena, and she does not reply, but when they look at each other their eyes shine like the surface of the Shimmer. This film repeats much of the theme of Stoker's novel in that the ecosystem around the Count's castle sends forth a representative, Dracula, to transform humans into creatures like him(it)self. Similarly, while in his own ecosystem he has a very different appearance from that which he assumes once he is out in the world, where he looks just like those around him—a doppelgänger of humanity, if not exactly mimicking a certain individual. As such, *Annihilation* can be read as the early stages of Dracula up until the point that the vampire leaves Transylvania,[39] except here the energy

---

39. Alex Garland's earlier film *Ex Machina* (2014) can also be seen to use the same section of Stoker's story as its narrative framework. See Bacon 2018, 225–33.

of the environment itself is concentrated within the "body" of the vampire, that is, Lena and her husband. Exactly how the eco-vampires will affect the world around them and whether it will become a "new" creation living under its own biologically excessive rules can only be guessed at, but as the film ends one thing is for certain: the vampire is not just connected or representative of the ecosystem that created it; it *is* that ecosystem.

CHAPTER 2

Vampiric Sustainability
The Undead Planet

This chapter focuses on how parts of the ecosystem take on more specific vampiric qualities to protect themselves and the wider environment from human incursion.

As touched upon earlier regarding *Dracula*, there is something of a synergy between vampires and bats. Stoker explicitly makes the connection by giving his large vampire bat the capacity to drain a human being—Lucy Westenra is sucked dry by Dracula in his bat form—though in the earlier part of the 19th century, this construction of the vampire bat was not as common as one might think. As noted by Kevin Dodd, the designation of "vampire often meant both the demonic being and the vampire bat—itself the object of much supernatural description until Darwin's recorded observations in the 1840s where the real vampire bat, *Desmodus rotundus*, is shown to be a type of micro-bat that only consumes relatively small amounts of blood."[40] Indeed, the wings given to vampire-like creatures more often denoted their demonic origins rather than signifying any kind of evolutionary connection with a real bat. Consequently, Stoker's text represents the most important (early) example of a vampire capable of becoming/transforming into a vampire bat and one that is intimately connected to the environment around the Count's castle. As mentioned in the previous chapter, vampires that are intimately connected to the ecosystem around them can take on the form of the various fauna of that environment. One could argue that, in being representative of that system, the vampire is actually constituted of those animals and creatures so that it not so much transforms into them but allows aspects of them to come to the fore—hence why Stoker's Count exists as a cloud of matter that coalesces into the various forms it wants to assume. The undead or undying ecosystems de-

---

40. Charles Darwin, *Narrative of the surveying voyages of His Majesty's Ships Adventure and Beagle between the years 1826 and 1836, describing their examination of the southern shores of South America, and the Beagle's circumnavigation of the globe. Journal and remarks. 1832–1836* (London: Henry Colburn, 1839), 3: 24–25.

scribed in this next selection of films similarly manifest themselves as varying types of lifeforms, including bats, as a means to defend themselves against the human invaders that have entered into, or threaten, their domain.

Vampire Bats, Eric Bross, 2005 / *Nightwing*, Arthur Hiller, 1979

By the late–20th century, the connection between vampire bats and vampirism is virtually a given, regardless of whether the bats shown are the size of the real thing—*Desmodus rotundus* is approximately 3.5 inches (8.9 cm) long with a wingspan of 7 inches (17.8 cm)—or human/bat hybrids, as seen in *The Bat People* (Jameson 1974).[41] An earlier instance in *The Devil Bat* (Yarborough 1940) is considered vampiric largely because its lead actor is Bela Lugosi— typecast in his role as Count Dracula even when playing a mad scientist, as seen in this film.[42] The bats in this film are huge, and although they attack and kill people by biting their necks, there is little else that would connect them to real vampire bats or even *Dracula*. More interestingly, the bats are scientifically altered, prefiguring the cultural anxieties around radiation and science in general that influenced many post–World War II films. Indeed, one of the first important postwar vampire narratives utilizes this concept to great effect. Richard Matheson's *I Am Legend* (1954) only obliquely mentions vampire bats, but they are pivotal to the story. The narrative tells of a world that has been overrun by a mysterious plague that turns everyone into vampires except for a man named Robert Neville. It transpires that he might be immune to the contagion due to his having been bitten by a vampire bat while stationed near the Panama Canal and that this has created antibodies that are resistant to this new plague. The novel suggests that the disease is somehow due to radiation, and its resultant mutational effects on bacteria and other forms of biological life further link it to folkloric and cultural beliefs connecting vampire bats and vampires. Furthermore, the disease is manmade, while its cure is nature(bat)-made.[43] Neither *I Am Legend* nor *The Devil Bat*

---

41. The bat people are not shown to be necessarily vampiric in the sense of needing human blood, but if they bite someone, the victim also becomes a bat person.
42. It should be noted that the sequel to the film, *Devil Bat's Daughter* (Wisbar 1946), more clearly references vampires.
43. The film *Daybreakers* (the Spierig brothers 2009) suggests a similar connection between bats and a global plague of vampirism, although the "cure" is very different from that in *I Am Legend*. *The Passage* by Justin Cronin and its television adaptation of the same name (Helden 2019–present) also use the idea of a natural property of vampire bats—in this case from the South American jungle—being mutated by man (the government) and creating a plague that turns humans into vampires. An interesting take on these themes is seen in *Chosen Survivors* (Roley 1974), which is set in the future and sees a group of chosen survivors being forcibly taken into a nuclear fallout shelter at the start of nuclear war. Once there, however, they discover that the shelter has been infiltrated by a colony of vampire bats that begin to kill the human inhabitants.

makes explicit the idea of human incursion or displacement caused by the destruction of a creature's natural habitat, but this idea is integral to the two films discussed below.

Vampire Bats proudly displays its ecological credentials, laying the blame for the invasion of deadly vampire bats into a small-town university campus in Louisiana on both human activity and illegal tipping of toxic waste materials. In this respect, the film is similar to a vein of horror films from the early 2000s, including *Cabin Fever* (Roth 2002), which specifically cites toxic waste as responsible for horrific and deadly mutations/diseases, and *Bats* (Morneau 1999) and *Bats: Human Harvest* (Dixon 2007), which see purposeful or accidental genetic modification to biological life as lethal to humanity. In fact, *Vampire Bats* is a sequel to the slightly earlier *Locusts* (Jackson 2005), which saw bioengineered insects threatening America itself such that only biology professor Maddy Rierdon can save the nation, a role she reprises in the later bat-centered film.

Bross' film follows its main protagonist Maddy (Lucy Lawless) as she becomes something of an eco-warrior battling against the human abuse of the environment when she takes up her post at the university. She becomes involved when some of her students, along with a few locals, are killed by having all the blood drained from them. She soon discovers that although the victims' bodies are covered in bat bites, they are not normal ones, as they have extra fangs allowing them to create bigger wounds to drink more blood. These genetic changes have also made the bats larger and dependent on greater sustenance than normal. In this sense, the bats act more like flying piranha fish that swarm over their victims, something that is further suggested by the fact that the bats originate from South America and that their nesting site is located near water. Water is given greater importance when it transpires that the local waste recycling plant is illegally dumping toxic chemicals into a nearby water supply, infecting the deer who are subsequently fed upon by the bats—the toxins seem to have far greater mutational effects on the bats than the deer.

The narrative is ambiguous in its reading of the behavior of the deadly bats, as at one point in the story Maddy suggests that it is human interference that causes the creatures to abandon their natural habitat to find a new home, which in itself is not seen as dangerous but simply an inevitable, if thoughtless, and undesirable consequence of human life. However, it is only once the bats arrive in America that they are turned into deadly killers.[44] This bad influence is blamed in part on human greed and corruption, and it is this aspect that is most closely tied to the vampires themselves, representing

---

[44]. This posits an interesting reading of Mexican and South American immigration to America.

them as voracious, insatiable, deadly, and appropriately feeding on the youth/future of the nation. In this respect, the deforestation or erosion of habitats by human incursion into the bats' original environment is marginalized, with the solution seen as local rather than global. As the story unfolds, it is quickly discovered that the mutated bats are the cause of the unexplained deaths, and the town's mayor calls in the state environmental services, which appear to be only one man. He plans to capture one of the bats and spread poison on it so that when it returns to its home the others will lick it and be poisoned—bats lick each other as a social practice. Maddy is not so keen on this idea and wants to understand them more clearly before attempting to tackle them, not least because the cause of their mutation is still unknown. Unsurprisingly, the environmental service's plan does not work, but Maddy starts to look more closely at what is going on with the help of some of her students, and they not only find out that the bats are drawn to a particular frequency of sound but that waste is being dumped in a nearby lake. She confronts the mayor, though it transpires he has been investigating the issue himself and has discovered that one man is responsible, the environmental services man, who has apparently been covering up information about the effects of the waste. This leads to the inevitable denouement where both the bats and the crooked environmental services agent are killed at the same time—lured into the steam vents under part of the university campus and boiled to death—seemingly resolving the problems of toxic waste and the deadly mutated bats in one foul swoop.

The bats are then an embodiment of the poisoned environment trying to rid itself of humanity and the greed and corruption that caused the poisoning in the first place. Consequently, their destruction is viewed as emblematic that all is resolved and life can continue as before. This offers rather simplistic solutions to real, ongoing ecological issues around the short-term accumulation of human wealth and the ongoing survival of the planet—a point of extreme contention in the first term of the presidency of Donald Trump, during which proposed environmental solutions are even more simplistic and self-serving than in *Vampire Bats*. The film *Nightwing*, however, takes a far more complex stance on very similar issues that are shown to be far more malevolent in their intentions.

Nightwing (Hiller 1979) is a more serious take on similar themes, although it draws some curious connections. The film itself comes in the wake of *Jaws* (Spielberg 1975) and a plethora of animal/eco-exploitation copies, such as *Grizzly* (Girdler 1976), *Food of the Gods* (Gordon 1976), and *Orca* (Anderson 1977), which showed nature to be intent upon the destruction of humanity—something of an interesting contrast to earlier narratives based on Wells' *War of the Worlds* (1897), in which the planet itself comes to mankind's aid. These movies are less environmentally or ecologically minded than

Vampire Bats, instead focusing on the precarious place of humanity on the Earth rather than true concern about the exploitation and poisoning of Mother Nature. However, *Nightwing* attempts to be more aware of such "green" issues and points toward both enforced migration due to human incursion and the destruction of habitats as well as the exploitation of nature through drilling for fossil fuels on sacred ground. These points become increasingly complicated as the film progresses.

The story is set on a Native American reservation in New Mexico that is shared by Maskai and Pahana peoples—in the novel of the same name by Martin Cruz Smith published in 1977, who also wrote the screenplay, it was the Hopi and the Navajo[45]—and where some mysterious animal deaths have occurred. A Maskai deputy, Youngman Duran (Nick Mancuso), and the chairman of the Pahana Tribal Council, Walter Chee (Stephen Macht), are called in to investigate the cause. Chee wants it all cleared away quickly because he is planning on selling large portions of the land to Peabody Oil, and more specifically the area of Maskai Canyon, which is sacred ground, so that the community, and mainly himself, can become rich off the profits. The impending desecration of such holy ground causes the Maskai shaman, Abner Tasupi (George Clutesi), to cast a spell to bring an end to the world. Abner's spell thus changes the meaning of the swarm of bats from only representing the just rewards of humanity for deforestation in South America to holy destiny for environmental desecration. This last is somewhat bolstered by the correlation in the film between ecology and the historical Native Americans so that the Maskai people who respect their tribal past, such as Abner and eventually Youngman Duran, are represented as being intimately connected to the ecosystem.

The film insists on emphasizing the "vampire" part of the eco-threat and overlays aspects of *Dracula*, introducing a Van Helsing–style vampire hunter named Philip Payne (David Warner) who has tracked the colony and their trail of dead from Brazil. He also overdramatizes the danger represented by the swarm, seeing them as catastrophic in nature, purposely equating them to flying vermin and the rats that followed Count Orlok in *Nosferatu* that spread bubonic plague in their wake.

This connection to the historical and cinematic vampire is made in an almost hysterical speech Payne delivers to Youngman Duran when the Maskai asks him about killing bats:

> Not just bats. Vampire bats. I kill them because they're evil. There's a mutual grace and violence in all forms of nature; and each species of life gives something in return for its own existence. All but one. The freak. The vampire bat alone is that species.... I killed over 60,000 of them last year in Mexico. You really understand the presence

---

45. See Ashlin 2019.

of evil when you go into their caves. The smell of ammonia alone is enough to kill you. The floor of the cave is a foul syrup of digested blood. And the bats: up high, hanging upside down, rustling, fighting, mating, sending constant messages, waiting for the light to fade, hungry for blood, coaxing the big females to wake up and flex their nightwings to lead the colony out across the land, homing in on any living thing; cattle, sheep, dogs, children, anything with warm blood. And they feast, drinking the blood and pissing ammonia.[46] I kill them because they're the quintessence of evil. To me, nothing else exists. The destruction of vampire bats is what I live for [Hiller 1979].

Up to this point, the deaths have been configured largely as an almost natural, if unusual, phenomenon—a larger than usual amount of vampire bats brought together because they were displaced from their habitats by humans. The above quotation suggests, however, that Payne sees this as a much more sinister, even Satanic, occurrence that is not accidental but has a specific target in mind.

This configures the bats' arrival in Maskai Canyon as the intentional corruption of holy ground, just as the oilmen are also threatening it. Abner's call to the death god equally increases the bats' credentials as representative of the "dark lord" and a cleansing force of dark ecology, manifesting what John Edgar Browning describes as the "apocryphal moment" (Browning 2015, 107). It is almost surprising that Payne does not carry a bible and silver crucifixes with him as part of his vampire bat–hunting paraphernalia. Indeed, something of this point is made when a group of Christians travel through the reservation and are attacked and killed by the bats, simultaneously showing the animals' antipathy to Western religion but also the connection to a different kind of guiding influence, not from above but from the land around them. Browning rightly connects this to the forces of modernism that are set as oppositional to tradition and the past (Browning 2015, 101), where the former is seen as all-consuming self-interest and the latter is founded on a kind of symbiotic relationship with the environment.

Unlike Bross' film, *Nightwing* does not show the bats as being changed by the land they end up in but rather as becoming a tool of that environment to rid itself of those who wish to exploit it; Maskai Canyon utilizes the arrival of the bats to prevent the drilling company from extracting its oil reserves. Within this is the ancestral connection between the land and the Maskai people, suggesting that all those who do not embrace their past and tribal traditions will be counted as a threat to the ecosystem. Consequently, it is only Youngman Duran who can save the surrounding population from Abner's wrath and the consequent apocalypse, as Duran is the closest thing that the shaman had to a son. Unsurprisingly, at the film's end, the vampire bats have

---

46. This is a feature of the vampires used in Guillermo del Toro and Chuck Hogan's *The Strain* trilogy of books (2009–11) and television series (2017–17) in which the undead creatures urinate while they feed.

The modern-day vampire hunters enter the vampires' lair. Kathryn Harrold as Anne Dillon (bottom left), Nick Mancuso as Youngman Duran (bottom, crouching), and David Warner as Phillip Payne (top) in *Nightwing*. Directed by Arthur Hiller (Culver City: Columbia Pictures, 1979).

chosen to roost in the most sacred cave within the canyon, the one that was formerly home to the original Maskai.

The cave also happens to be the point where the underground oil reserves bubble to the surface in large pools. The cavern becomes something of a vampire's lair, not unlike Dracula's castle, where the ruins of history hang heavily in the air along with "piles of [black] gold," creating a space that is simultaneously otherworldly and the focus of the environment around it. As with the Count's lair, the hunters—Youngman Duran, his partner and local nurse Anne Dillon, and Philip Payne—enter it during daylight hours, knowing they only have until nightfall to kill the vampires that cover the cave ceiling. Youngman Duran, as Van Helsing did before him with silver, crosses, and wolfsbane, places his trust in the ancient ways, donning a headband, baring his chest, and consuming the sacred Datura root—a hallucinogenic plant.[47] Youngman and Dillon enter the cave safely through a hole in the cavern roof, but Payne is entangled by the ropes hanging from it. The sun is beginning to set, and the bats are stirring just as Duran starts seeing visions of his ancestors and remembers the final paintings of Abner depicting the apocalypse. These paintings remind him of a circle of black stones—one assumes these to be solidified oil—which he begins to lay out on the floor using the stones that litter the floor of the cave. Payne manages to drop to the cavern floor but hurts his leg, and Dillon drags him to safety as the bats begin swarming around them. Meanwhile, Youngman is struggling with his visions as he sets fire to the stones, igniting the oil and filling the cavern with the flames that erupt from it, creating poisonous smoke that kills the bats. Furthermore, the fires render the pools of oil unusable and/or uncollectable for a very long time. This ending is very similar to that of *Vampire Bats* in that it destroys both the monstrous representatives of nature and the threat of capitalism and modernity, allowing for a sense of balance to return to the local ecosystem. The triumph of the old ways of interdependence rather than exploitation of the landscape has saved the Maskai from "the end of the world." Neither film infers that the evils of exploitation will ever happen again. This is not the case with the movies discussed below, in which nature utilizes far more invasive and unexpected means to discourage humanity and the threat they pose to the natural environment.

Surviving Evil, Terence Daw, 2009 / *Splinter*, Toby Wilkins, 2008

Daw's film marks a return to the trope of humankind, though mainly Western culture, venturing into areas that should be left alone, not unlike *The Black Water Vampire* but with a more obvious link to historical colonialism. The film implicitly contains the idea that ecological trauma is not instantly

---

47. Youngman's wardrobe at this point seems more symbolic for a 20th-century Western audience familiar with the Western genre than meaningful within the terms of the narrative itself.

cured or healed but, not unlike issues surrounding post-traumatic psychology,[48] is ongoing and recurring. The story sees a small film crew arrive on a remote island, Mayaman, in the Philippine archipelago to make an episode of *Surviving the Wilderness*, an ongoing series starring celebrity survivalist Sebastian Beazley (Billy Zane). The crew of six make camp in what seems to be an island paradise; however, it is not long before strange events begin to make them all uneasy. The next day, Joey (Joel Torre), who acts as the local guide, and Cecilia (Natalie Mendoza), who is of Filipino heritage, travel further inland to look for the village of the Isarog tribe, but when they get there, everyone is dead and a pregnant woman has had her stomach ripped open. Joey swears Cecilia to secrecy, but that night another one of the crew, Phoebe (Christina Cole), overhears them talking about the Aswang, a mythical creature of the Philippines.[49]

The name Aswang covers a diverse group of evil, shapeshifting spirits that have appeared in popular myths in the islands of the Philippines since at least the 16th century, though anthropologists have speculated that Spanish colonists invented/utilized them to control outlying areas of the local population.[50] In their various incarnations, the Aswang sometimes share many common features with what Western culture would see as vampiric or vampire-like, though in Daw's film they are ferocious tree-dwelling, shapeshifting creatures that desire blood and are drawn to pregnant women, as they seek to impregnate the fetus in an evolutionary process called "Dungo Nan bunti"—no explanation of this process is provided. The Aswang's only weaknesses seem to be a fear of fire and of underground spaces, though at various points they also seem to prefer darkness and shade to sunlight. Unsurprisingly, the creatures seem to be part of the interior forest of the island, appearing from nowhere and disappearing at will.

It is not long before Sebastian is bitten and Phoebe discovers the dead baby that was torn from the mother's womb in a tree near the beach. It transpires that Joey's grandfather had been held prisoner on the island by the Japanese during World War II and knew about gold buried under the camp by his captors. Following a map left by his relative, Joey and Cecilia, unfazed by the earlier discoveries, return to the village, where he blows up the floor of a gun tower not far from the Aswang nest to discover the hidden treasure.

---

48. See Georgiana Banita 2015, 146.
49. The Aswang form the basis of many stories and films. See Maximo D. Ramos' *The Aswang Complex in Philippine Folklore* (South Carolina: CreateSpace Independent Publishing Platform, 1971), *Aswang* (Martin and Poltermann 1994), *Aswang* (Tarog 2011), and *Aswang* (Laurin 2018).
50. See Nadeau 2011. Aswang are also historically connected to the idea of witches, as seen in Eastern Europe (Summers 2013, 58), Africa (Melton 2011, 5), and the Caribbean (Bane 2017, 97).

Cecilia, who is still above ground, is attacked by the Aswang and is only saved by falling into the hole created by the explosion. Joey tries to escape but is caught by the creatures and killed. Cecilia manages to get out from the hole and run back to the camp, but the creatures follow her and attack the surviving members of the crew. After much to-ing and fro-ing, two of the group find themselves back in the hole with the gold and a large amount of left-behind explosives. Trapped with no way to escape, they set off the charges, killing themselves and many of the creatures. The attack further increases the many correlations to Western vampires that occur throughout the film, where the Aswang are able to fly, bat-like, from their nests and take on many aspects of the fauna of the forest around them: climbing like monkeys, flying like birds, and even taking on the shapeshifting/nightmarish qualities of the jungle itself. Meanwhile, Phoebe fights off some Aswang on the beach—one of these is one of the crew who has mutated into one of the creatures—and manages to escape on the raft the group kept for emergencies. She loses consciousness and awakens once the raft has landed on another, presumably nearby, island and is taken in by some women to a hut to recover. Phoebe then discovers that she is surrounded by other pregnant women, all of whom are about to give birth to more of the Aswang.

There is much here that is not dissimilar to *The Black Water Vampire*, especially the use of human women to procreate and the seeming collusion of the locals, but the assimilation/mutation of outsiders suggests that there is more at play. The film suggests that the Aswang were created by the violence of the Spanish colonizers and that the cruelty and savagery that was inflicted on the local population and the environment was somehow made manifest in these supernatural creatures. This premise forms something of a link to *Nightwing*, which rather loosely alludes to ancient prophecy/magic bringing the bats forth. In this sense, newcomers to the environment are themselves colonized, consumed, by the island, turning them from invaders into part of the indigenous ecology. The Aswang become active agents in this scenario, manifesting a form of environmental antibody that attacks foreign invaders but goes further than antibodies by converting the outsiders and adding them to their number, either by mutation or reproduction. The implication is that the Aswang themselves are predominantly male, or at the very least unable to procreate among themselves. Unlike in the previous two films, the ending is more ambiguous. Although Phoebe is obviously going to give birth to another Aswang, it is unclear whether the offspring delivered in the "maternity" room on the other island are returned to Mayaman or will inhabit this new home. The question of whether the environment's intention is only the integration of outsiders into the localized ecosystem or whether it is intent on spreading further remains open. The next film is far less ambiguous in its ending and the intentions of the aggressive and excessive forces of nature that want to consume not just humanity but all living creatures.

Splinter does not provide the comfort of a distant paradisiacal island or the psychological buffer of perceived unreality provided by the cultural manifestations of the supernatural or the folkloric. Rather, the film shows the uncontainable jouissance of the biological beyond the anthropocentric. Set in an unnamed forest in middle-America—it was actually filmed outside of Oklahoma City—the film begins and then centers around a small service station in the middle of nowhere surrounded by woodlands. As the film begins, the sole attendant of the station is seen sitting outside on a folding chair, enjoying the sunshine. Above the almost deafening sound of the crickets and insects chirping, he hears something rustling in the grass behind him. He starts to walk through the ankle-high grass to find out what is making the noise when he is suddenly attacked by something that looks like spike-covered roadkill.

The screen fades to black, and the scene cuts to a young couple, Polly (Jill Wagner) and Seth (Paulo Costanzo), who are camping in the woods but, upon damaging their tent, decide to spend the night at a motel. They stop to help a girl by the roadside, Lacey (Rachel Kurbs), only for her boyfriend, Dennis (Shea Whigham), to appear by their car holding a gun. As Polly and Seth discover, Lacey and Dennis are criminals on the run whose car has broken down and who are therefore in need of the couple to hitch a lift to the nearest town. As they are driving, night falls and the car suddenly hits something on the road that punctures one of the tires. They all exit the car, and Lacey and Seth go to see what they hit while Dennis and Polly change the wheel. The tire is covered in some gooey substance, and Dennis manages to get a splinter of some sort in his finger. Meanwhile, Lacey asks Seth to pick up the dead creature—something similar to a small deer or a dog—but as he bends down, parts of it seem to move of their own volition. Spooked, Lacey runs back to the car and they continue their journey, but the engine overheats, and they are forced to pull in to the first garage they come across, which just happens to be the one seen at the beginning of the film. All the lights are turned on in the station, but no one is about, so Dennis takes Polly into the shop to procure drinks and snacks while Seth refuels the car and sorts out the engine. Lacey needs to use the bathroom and therefore circles around the side of the building, but the door appears to be blocked. She finally manages to get the door open, only to find the attendant slumped on the floor, covered in blood with three-inch-long needle-like spikes protruding from his body. Lacey staggers back and runs around to the front of the station, looking through the window at Dennis as the body seems to launch itself out of the room, not quite running or in control of itself, and toward the girl, knocking her to the floor and banging her head with extreme force. The spine-covered figure of the attendant crunches onto the hood of the group's car and remains

Chapter 2. Vampiric Sustainability

Lacey's hand takes on a life of its own in *Splinter*. **Directed by Toby Wilkins (New York City: Magnolia Pictures, 2008).**

there with Lacey's body prone on the floor nearby. Seth had run into the service shop as soon as the body appeared and stands with Polly while Dennis goes to the shop door, assessing whether he can reach his girlfriend before the body on the car would get to him. He sneaks out and tries to take hold of his girlfriend, but, unseen by him, her hand begins to search for him. As he realizes this and starts to run back to the shop, the girl's body launches itself at him. As Dennis slams the door shut, Lacey's hand breaks off, grows spines out of the stump and starts scuttling across the floor, seeking them out, seemingly led by the warmth of their bodies.

They drop a heavy bag on top of the hand, crushing it, and although it is unable to move afterward, a black ooze comes out of it that seems to be trying to regenerate itself.

Meanwhile, the figure on the car and Lacey seem to have combined in some way, and the newly formed "body" begins throwing itself at the shop door, drawn by the warmth of the humans within. They hide in a room at the back of the shop, but the figure climbs onto the roof and attempts to find a way in. They hear a voice outside and rush back to the front of the shop where a policewoman has arrived, attracted by the blood on the station forecourt. The officer (Laurel Whitsett) sees the gun in Dennis' hand and demands he drop it, but he will neither comply nor allow her to approach, and she grows increasingly frantic even though Seth and Polly are screaming for her to get back in her car. Suddenly, from above, the mutated figure grabs the officer and pulls her in two, taking the top half up onto the roof and joining it to the creature's own body.

This process is key to assessing the nature of the entity. The creature drags the top half of the body onto the roof, and spines/tendrils reach out from its main mass—it does not have a conventionally organized body and adds new pieces to itself at random—that latch onto the flesh of the torn-open torso, pulling it into the mass, simultaneously making it part of the whole and mutating its cells so that the new addendum sprouts spines. This graphically shows how the cells of the new entity are all equally energized with no specific job or identity within the whole—i.e., none are allocated as eyes, legs, internal organs, etc., but they become an amorphous part of a living mass. Thus, Lacey's hand moves on its own accord, and the new incarnation of the creature similarly loses a larger part of an "arm" that hunts the survivors inside the shop but uses all parts of its mass to move—this is very different from the more traditional representation of reanimated limbs as demonstrated by "Thing" in *The Addams Family* (Levy 1964–6 and Sonnenfeld 1991). Indeed, the creature here bears more resemblance to the alien entity in *The Thing* (Carpenter 1982) in its propensity to assimilate human form and energize every part of itself, not unlike Newsom's earlier comment about ecological jouissance without utility other than being increasingly itself. However, the *Splinter* creature does not copy human form by consuming part of it but instead integrates the form into itself by mutating new cells to be like its own. Thus, it constructs its vampiric nature in the way it consumes and converts biological material into itself, gaining life from life. It is not unlike the unstoppable bioenergy seen previously in *Annihilation*, which exceeds all known forms of human/animal categorization to become an expression of biological life itself or a manifestation of the environment integrating all life back into its ecosystem.

Meanwhile, the wound in Dennis' finger has been getting progressively worse, with the black veins that flowered out of the original tiny splinter now controlling his arm almost up to the elbow. The energizing quality of the virus/infection is shown as the flesh of his hand disregards the skeleton beneath it and bends his fingers and wrist in extreme and violent spasms, audibly breaking the bones beneath. To try and save him, the survivors decide to cut off his arm just below the shoulder—a scene not uncommon in zombie films. They successfully manage this and immediately dispose of the severed limb. During their earlier tussle with the arm of the creature, the survivors, while hiding in a freezer-cabinet in the shop, discover that it responds to and seeks out heat—equating heat with biological life—surmising that if they lose enough body heat they become invisible to it. Seth volunteers and cools himself enough to barely function and reach the car parked outside, but he incvitably attracts the attention of the larger creature and the severed arm—it is a bit misleading saying "arm," as it is really a smaller creature in its own right—in the shop, both of which begin to move toward him. Dennis runs out of the backroom with a gun and shoots the arm and one of the fuel pumps, causing it to explode in flames all over the larger creature. Polly runs to join Seth, but Dennis waves them off as he shoots a stack of gas canisters stored near the shop, making the entire station erupt and destroying all parts of the creature. As Polly and Seth drive away, the camera follows them and hovers high above the car as it disappears along the long, straight road through the woodlands.

The camera then pans around back into the trees to a large trunk that is oozing black, sticky fluid and which appears to have some life of its own as it pools, awaiting a new victim. This final shot links the creature, or fungus as many reviews describe it,[51] to its environment and the larger ecosystem. Unlike *The Thing* or *The Hallow* (Hardy 2015, described later), it does not need human incursion to be aroused but is actively venturing out into the human world. This ending also clearly shows that this was an opening skirmish with more to follow. There is no ambiguity here but a clear affirmation that the reclaiming of the planet and the reincorporation of humanity into the biological mass of the ecosystem are inevitable. This naturally leads to the idea of the environment as a discrete entity taking measures to ensure its own ongoing protection and regeneration rather than manifesting or directing creatures/entities to do this on its behalf, thereby creating what might be called an undead environment.

Burnt Offerings, Dan Curtis, 1976 / *The Pit*, also known as *Jug Face*, Chad Crawford Kinkle, 2013

As discussed above, landscapes or places that produce vampiric entities are not uncommon. Indeed, one can argue that Count Dracula himself is a product of his historical and geographical location of Transylvania, a land frequently afflicted by the trauma of conflict and bloodshed that almost seems to bring forth a creature born, or manifested, from that violence and its concomitant desire for human blood and suffering. Coming from this, undead environments are themselves vampires, whether produced by past traumas that occurred in them or by an entity that exists within them—this latter not unlike Stoker's own tale *The Lair of the White Worm* (1911) and its later adaptation to film by Ken Russell in 1988. Algernon Blackwood's "The Transfer" (1911) is an interesting example in which a small patch of land at the

---

51. The synopses of the film on IMDb, Filmweb, and Wikipedia all call the organic mass a fungus.

bottom of the garden of a house in the countryside drains the energy of all those who approach it. There is no backstory provided as to why the piece of ground is the way it is, but the owners of the house are aware of its mysterious power. One day, a particularly obnoxious relation of the owner's visits, and they allow him to be "caught" by the lure of the piece of ground: "he stepped forward into the middle of the patch and fell heavily upon his face. His eyes, as he dropped, faded shockingly, and across the countenance was written plainly what I can only call an expression of destruction" (Blackwood 2011, 5). Afterward, the victim, though not dead, falls out of public life and is not heard of again, while the patch of land bursts with life, "full of great, luscious, driving weeds and creepers, very strong, full—fed, and bursting thick with life" (Blackwood 2011, 6).

Burnt Offerings contains much of the spirit of "The Transfer," not least in that the vampiric environment is a garden, though it also includes the house that stands upon it. Unlike Blackwood's story, the garden requires energy, particularly blood, in order to restore itself to its former condition and a more balanced ecology. The story begins with the Rolf family—Ben (Oliver Reed), Marion (Karen Black), and their son, David (Lee Montgomery)—arriving at an old, rundown Victorian mansion in a remote part of the countryside that they are planning on renting for the summer. The film includes a long scenic drive, suggesting a journey to another land/world, as was later seen in the opening sequence of Stanley Kubrick's film *The Shining* (1980), which also speaks of an undead environment and historical trauma. The family is met by Arnold and Roz Allardyce (Burgess Meredith and Eileen Heckert), the idiosyncratic brother and sister who own the house and look after their aged mother in its top room. The rent for the summer is extremely low, but it entails the condition that their mother remains undisturbed upstairs and that she is left a meal in her room at particular times each day. Marion is extremely keen to rent and persuades Ben to agree to the terms. Neither notice their son outside playing and cutting his hand as he falls out of a tree in the garden, nor that after his fall, one of the formerly dead plants in the observatory is suddenly sprouting a green leaf. The Allardyces are very excited by this and even more so when they find out that the Rolfs will be bringing an elderly aunt with them and reassure the couple that "the house will take care of itself." A deal is struck, so the Rolfs leave to collect their belongings and Aunt Elizabeth (Bette Davis) and return a few days later, but no one is there, and they find only a note from the Allardyces welcoming them to the house.

The brother and sister fulfill a role similar to Renfield's for the property as a vampire's assistant who lures unsuspecting prey into its grasp. In fact, there seems to be no sign that the siblings actually live in the house, as though they might be a projection created by the property itself with the purpose of bringing humans into its reach. Once settled, the Rolfs begin to explore the house and its grounds, but it is not long before strange things begin to occur, not least in changes to the behavior of both Ben and Marion, who become increasingly aggressive to, and distant from, each other. This effect resembles that of Dracula upon his victims in Stoker's novel, and similarly, once the property begins to draw energy from its new occupants, it starts to forge a psychological link to them, particularly Marion. In this sense, she is a Mina Harker figure in the story as the character who is closest to the "vampire" and effectively becoming one with it. Marion was instantly drawn to the property, and once they move in, she sets about cleaning and polishing the interior to the extent that she seems to put the property above her own family. She equally becomes devoted to the Allardyces' mother, who never seems to see or hear her, but Marion leaves and removes trays of prepared food for her every day. The old lady's room is divided into two sections: one a bedroom, which we never see, and the other a living room of sorts with a chair in front of a window that is the topmost in the house and functions as an all-seeing eye over the property. The interior is immaculate, and Marion is fascinated with the many framed photographs that fill the top of a table. They show people from all eras, at least as far back as the start of the 20th century if not further, but with no suggestion of who they are or how they are connected to the Allardyces. Things come to a head a few days later when the clocks in the house, which had all stopped, suddenly reset and start at midnight. This wakes Ben, who then smells gas and traces it to David's room, but the door is locked. He breaks it down to rescue his son, but no one knows why the gas infiltrated the room in the first place. In the morning, Elizabeth, who has suddenly become dithery and frail, confesses she was in David's room but cannot remember if she touched the gas fire or not. Marion accuses her of sabotage, and the old lady runs off to her room, crying. Ben goes to see her to apologize on behalf of his wife and urge reconciliation, but Elizabeth is too weary to move. She falls asleep, and when she awakes and tries to move there is an audible cracking of bones, as though they had been sucked dry and become brittle. David runs to fetch his father, and Ben calls for a doctor while Marion sits in Mrs. Allardyce's room, eating the prepared meal and listening to the music box sitting among the many photographs on display. Later, we see Marion enter the conservatory downstairs, where all the previously dead flowers are in full bloom.

After Elizabeth's funeral, Ben tells Marion that they are leaving the house. She objects that she cannot leave the old lady, but her husband insists. That night, however, Ben is awakened by the sound of bricks and tiles falling to the ground; the house is quite literally shedding its old skin to be replaced by a new one. Panicking, Ben grabs his son and runs down to his car to leave immediately, but the weather grows rapidly worse, with heavy winds and pouring rain. As the car starts driving on the narrow road leading out of the property, trees fall in its path, and Ben tries to ram them out of the way. Failing, he leaves the car but is hit in the face by branches from the trees, and tendrils of ivy curl around his legs, pulling him to the ground. Marion arrives and drives them all back to the house, but her husband has suffered some kind of seizure that has left him paralyzed. The following day, David is swimming in the pool next to the house when the water becomes increasingly choppy, dragging him under the surface. Marion can see this from within the house, but the door to the room she is in locks itself, and she must smash a window to escape and save her son. During this time, Ben has managed to struggle out of the chair he was in and somehow "broken" his seizure. The following morning, Marion loads them all in the car to leave but decides she should quickly tell Mrs. Allardyce they are going. Ben and David wait for her, and after 10 minutes she still has not returned. Ben goes up to the room to see what is taking so long. As he enters, he sees the back of a gray-haired lady wearing Victorian clothes sitting in the chair by the window, and as he swings the chair around, he realizes it is Marion herself. Suddenly, we see Ben flying through the window and landing on the car below, terrifying David, who is still inside it. Before the boy can react, the old chimney of the house falls off and crushes the car, killing David.

The film concludes with the Allardyce siblings returning to the house, which now looks brand new with beautiful green gardens and flowers in full bloom. Arnold looks admiringly at his surroundings and says that their mother has been restored to her full beauty, which again points back to the idea that they are a projection of the property in some way, the lure for the Venus flytrap of the house and gardens to feed off. Within this is the notion that the house has not just been made new but has returned to its past—the actual history of the house is never mentioned, but the array of photographs of previous victims displayed in the upstairs room suggests that it is of the late Victorian era—and it has fed upon modernity to go back to a time when its micro-ecosystem was balanced and fecund without human interference; the gardens seem to require no one to keep them trimmed and in bloom, though perhaps this is also done by the "children" of the property, Arnold and Roz.[52]

The next film shifts this slightly and to some extent refers more strongly to Blackwood's tale mentioned above, showing the vampiric effects of a hole in the ground that also controls the environment around it.

Jug Face, more aptly also known as *The Pit*, is about a largish hole about

---

52. The series *The Haunting of Hill House* (Flanagan 2018) suggests a similar scenario, though in *Burnt Offerings* the haunting is limited to just one room rather than the property as a whole. Here, the "Red Room" in the house appears in different forms to the members of a family that live there, specifically playing on their individual fears so that it can feed off their psychic energy. H.G. Wells also wrote a short story called "The Red Room" (1894) that describes something similar about a room that is haunted by fear.

The Allardyces' "mother" before (top) and after. *Burnt Offerings*, directed by Dan Curtis (Beverley Hills: United Artists, 1976).

A vampiric hole in the ground in *Jug Face*. Directed by Chad Crawford Kinkle (Los Angeles: Moderciné, 2013).

10 feet across that has thick, muddy, often bloody water at the bottom. Its underwater depths are never shown, but it is about 10 feet from the surface of the water up to the ground above, and its sides are vertical with broken roots and such protruding from them.

Some background is provided in an animation sequence at the start of the film where, in childlike drawings, we are shown the hole in a clearing surrounded by forest. From the clearing, a path leads to a church and a few houses. A couple are staying in one of the houses, dressed in 17th-century attire and accompanied by a child. The wife is very ill and stays in bed. The husband experiences a vision of some sort and then goes to a barn/workshop where he molds a pot with the image of a human face on it. The face is that of the church leader, who is taken by the townspeople to the clearing, where his throat is cut so the blood flows into the hole. The man's wife has suddenly recovered her health. The film then begins in earnest, and we are in a present-day backwoods community. A girl, Ada (Lauren Ashley Carter), is being chased by a boy, Jessaby (Daniel Manche), who is soon revealed to be her brother. He catches Ada, and they have sex near the pit, which begins to ripple slightly and causes a man, Dawai (Sean Bridgers), in the nearby camp to go into a trance and begin creating a pot. Unaware of his actions throughout, he sculpts a face in a jug, takes the jug and places it in the kiln in his yard, and returns to his shack and falls asleep. Later that day, Ada goes to see Dawai and looks in the kiln and discovers the jug has her face on it. Knowing that Dawai has not looked at it yet, she takes it and buries it in the woods.

The community is extremely small, made up of only five or six families who purposely keep their distance from the outside world. Only Sustin (Larry Fessenden), Ada's father, goes to the nearest town to trade, and on his next trip he takes his daughter with him. While in town, Ada manages to steal a pregnancy testing strip and later discovers she is pregnant with her brother's baby. Back at the community, Ada is in the river washing clothes with Eileen, the sister of Ada's betrothed. Suddenly, Ada falls into a trance where she sees Eileen killed by a mysterious creature, and as she comes to she rushes to the pit to find her friend torn to pieces around its edge. This immediately arouses the suspicions of the community, as no one ever dies unless something has interrupted the ritual process they live by—the pit causes Dawai to go into a trance and carve the face of the required sacrifice onto a jug, and that person has their throat slit over the hole.[53] The villagers go to Dawai to see if he is missing a jug, but he denies remembering anything, and so they insist he look everywhere just in case. Ada goes to see him later and suggests he try to re-sculpt the jug from memory. The townspeople return the following day, and he presents them with a pot with the face of Ada's betrothed on it, and he is dutifully killed.

Ada visits her sick grandfather and sees a vision of an emancipated boy who tells her that her grandfather was punished by the community for doing exactly what she has done when he hid a jug bearing his wife's face. The boy appears to be linked to the pit, as he made a previous appearance when Eileen was killed, but whether he is a positive or negative aspect of the pit's nature remains unclear. Jessaby tells his father he is not feeling well, so Sustin takes him to bathe in the pit. All seems fine until Ada, who is nearby, falls into a trance and sees her brother killed. Sustin begins to pull his son from the pit, but he is suddenly grabbed from below. Blood spurts up out of the hole and the boy is gone. The community realizes something is wrong and chains Dawai to a fallen tree near the pit so he can be taken by it, but Ada releases him later that day, and they both run away. Unfortunately, they are caught, and it is assumed that the two are in a relationship, so they are whipped. This causes Ada to miscarry, and she reveals that it is her brother Jessaby's baby. She then falls into another trance, and this time her father is killed. Both she and Dawai are tied up near the pit, but her grandfather and the boy arrive to release her. Ada decides she must give the pit what it needs and save Dawai, so when the townspeople return the following morning, she is waiting by the hole for them. She is duly sacrificed, breaking the cycle and satisfying the desire of the pit.

---

53. It is never explained if the residents are allowed by the pit to live until they die of natural causes or whether they are inevitably called for as a sacrifice before this can occur.

There appears to be a creature of sorts in the pit beneath the water, and when members of the community go into a trance, we are shown the pit's surface bubbling or rippling, but nothing further is revealed. All we really see are the effects the pit has on those in its vicinity. Its behavior resembles Dracula's in its psychic connection to the minds of those who are its intended "victims." It is never made clear why the community is still so connected to the pit, as their worship does not quite account for the intensity of their thrall. They appear to be hostages to its influence, and it appears to require their blood to survive.[54] Consequently, it acts as something of an environmental regulator, creating and controlling the area around it—a quite substantial area given how distant the community is from a large populated town—and limiting the effects of the human presence within it. All the houses/shacks that the villagers live in look makeshift and surrounded by salvaged materials that they obviously reuse to patch and repair their dwellings. In addition, there only seems to be one working truck, driven by Sustin. The community hunts for food or eats roadkill—Ada's father runs over an opossum and takes it home with him. They thereby live in balance with their environment and are happy to sacrifice one of their own when the time comes—all but Ada's grandfather, of course, who decides to spare his wife instead. In a sense, the pit acts as a kind of black hole that sucks in time, stopping the community from entering, or falling subject to, the advancing time of the world around it. Again, not unlike the Allardyces' "mother," the hole and the environment around it is in a sense undead or immortal, never changing but using the energy (blood) of humans to sustain itself.

The next films under discussion continue these ideas but in more physically and traditionally vampiric ways.

Dracula Untold, Gary Shore, 2014 / *Primal*, Josh Reed, 2010

More than the previous examples, the two films discussed in this section define the source of environmental reaction and self-preservation as explicitly produced or energized by a central figure/entity that aggressively stymies human incursion, whether to maintain its own integrity, as in *Dracula Untold*, or to provide sustenance for its continued existence, as in *Primal*. *Dracula Untold* is one of the many vampire films that works on the premise

---

54. In some ways, the idea used here is picked up in the more recent film *The Ritual* (Bruckner 2017), which also takes place in a remote woodland/forest and features an ancient creature that keeps a community in thrall around itself to worship it and feed upon, though here it is fear and human pain that act as sustenance.

Chapter 2. Vampiric Sustainability 69

that Vlad Țepeș, more popularly known as Vlad the Impaler, is none other than Count Dracula as invented by Bram Stoker—Stoker certainly knew of the historical figure but did not explicitly make the connection.[55] Vlad III (1428/31–1476/77), a.k.a. Vlad Țepeș, a.k.a. Vlad Dracula, a.k.a. Vlad the Impaler, was an interesting character caught in a time of constant struggle between the Hapsburgs and the Ottoman Turks, with the borders between the two empires advancing and retreating across Eastern Europe and what is now modern-day Romania. Vlad and his brother Radu were taken as hostages by Suleiman the Magnificent, leader of the Ottomans, to ensure the loyalty of their father, Vlad II—known as Vlad Dracul (Dracul meaning Dragon or Devil), hence his son being known as Dracula (the diminutive meaning "son of the dragon")—who was in service to the Holy Roman Empire. Vlad's father and brother were murdered by John Huyadi, regent-governor of Hungary, who installed Vladislav II (Vlad's cousin) as voivode (prince) of Wallachia. Vlad began a series of campaigns to take control of Wallachia, enlisting the help of the Turks and later the Hungarians, with whom he managed to take control and maintain a six-year reign. He began a purge against Wallachian Boyers and the Transylvanian Saxons to strengthen his grip on power and impaled the latter en masse, which led to his nickname/cognomen.[56] The Ottoman Sultan, Mehmed II, demanded tribute from Vlad, but the Wallachian impaled his envoys instead. In 1462, Vlad launched an attack on Ottoman territory, killing tens of thousands of enemy soldiers. Unsurprisingly, Mehmed launched a reprisal attack on Wallachia, hoping to replace Vlad with his younger brother Radu, but after a failed attempt at capturing the Sultan, the Ottoman ruler left the region with his main army. However, the rot had set in his own ranks, and with his own Wallachians deserting him, Vlad went to Matthias Corvinus, the king of Hungary, in Transylvania but was imprisoned in late 1462. He was finally released in 1475 and was killed fighting for Corvinus' army in late 1476 or very early 1477. One other point of note in relation to

---

55. See Elizabeth Miller, *Dracula: Sense and Nonsense* (Desert Island Books, 2012), among others. Recently, however, Dacre Stoker, the great-grandnephew of Bram, has claimed that his ancestor knew much more about the historical Vlad than is currently provable; see Wayne Miller, "More on That Book Proving Bram Stoker Knew About The Historical Dracula," vampires.com, https://www.vampires.com/more-on-that-book-proving-bram-stoker-knew-about-the-historical-dracula/, accessed 9 June 2019. However, my colleague Andrew M. Boylan takes issue with this claim; see 'Will the real Count Dracula please stand up?' *Taliesin Meets the Vampire*, 27 November 2013, http://taliesinttlg.blogspot.com/2013/11/will-real-count-dracula-please-stand-up.html, accessed 22 August 2019.

56. It is worth noting that much of the negative press and fake news around Vlad III was printed and disseminated by the Saxon and Russian communities that were most affected by his control of Wallachia and subsequently manipulated and weaponized by the political intrigues of the times. See Squires, *The Telegraph*, 2010. Indeed, similar negative views surrounded one of his distant cousins, Erzsébet Báthory, who was equally caught up in the political machinations in the region. See Holloway, *Ancient Origins*, 2014.

Shore's film is that Vlad was married twice, with his second wife being Justyna Scilágyi, whom he married in 1475. His first wife might have been an illegitimate daughter of John Huyundi and carries the apocryphal tale of committing suicide when receiving false news of Vlad's capture by the Ottomans and their imminent storming of the castle/monastery where she was staying.[57]

Dracula Untold rather freely plays with historical facts to construct more of a troubled superhero figure—this is largely due to the film constituting an (ongoing) attempt by Universal Studios to turn their stable of classic monsters (Dracula, Frankenstein's Monster, the Mummy, and the Wolfman) into a franchise/universe not unlike that of Marvel or DC Comics. The action begins in Wallachia with Vlad (Luke Evans) as its prince, who commands the adoration and respect of his people. A voiceover narration at the opening of the film by Vlad's son, Ingeras (Art Parkinson), tells of dark times in the past and more to come in the future when he says, "My father was a great man, a hero, so they say. But sometimes the world doesn't need another hero, sometimes what it needs is a monster." Vlad himself talks of times in his past when he was a monster for the good of his people and their land but represses these memories, with only the scars on his skin as physical evidence they ever happened. Much is made of the idea of a king as integral to the health of a kingdom—with a kingdom viewed more as the entire ecosystem of people, land, flora and fauna, resources, etc.—suggesting that nature requires the right stewardship (often of royal blood), or it will wane and die.[58] Consequently, the film emphasizes the leader's connection to the land in its opening scenes in which he scouts the outlying regions of the kingdom and looks for signs of Ottoman incursion. He is leading a small party of men, and they spy some pieces of armor at the side of a stream. Vlad, who possesses intimate knowledge of his kingdom, can immediately read the signs and knows that something unusual caused the damage to the helmet. He consequently tells most of the party to return home. Those who remain follow the stream to its source at Broken Tooth Mountain and find a cave that is littered with bones. The unfamiliarity of the terrain configures it as a journey into the dark heart of the kingdom, and although Vlad's palace is the human center of the environment, this cave seems to be its reflection in a black mirror. If Harker's journey to Transylvania takes him out of his world and to a place beyond the forest, Vlad's takes him to its core.

As Vlad and his companions step into the cave, the companions are killed in short order, leaving him to face the evil within alone. Surprisingly,

---

57. This story is used in *Bram Stoker's Dracula* (Coppola 1992) and hangs around the demise of Vlad's wife in Shore's film.

58. This is seen in *Snow White and the Huntsman* (Sanders 2012) and in recent Korean media such as *Rampant* (Kim 2018), *Monstrum* (Huh 2018), and *Kingdom* (Kim 2019–present).

the monster (Charles Dance) allows him to leave, and Vlad returns to his castle, learning that the creature is the Vampire, a human who was tricked by a demon many years ago and is trapped in the cave. Meanwhile, the Easter celebrations are just about to begin, and an emissary of Mehmed II arrives, demanding a tribute of 1,000 boys to be trained as soldiers—a practice that had long been stopped but was now reintroduced by Mehmed. Vlad is distraught and unable to tolerate the thought of the future generation of his kingdom, his land, being taken away and corrupted in the same way that he was, but his wife (Sarah Gadon) is convinced that he can convince the Sultan to rescind his demands. When the appointed time comes, and Vlad offers himself in place of his son, the Sultan's representative refuses, prompting the Wallachian to kill him. Knowing the inevitable results of his actions, Vlad feels there is no option but to return to the cave and harness the dark power at the heart of his kingdom. The Vampire emerges from the gloom and confronts him about the consequences of his decision. The Vampire explains that once Vlad has drunk his blood he will have three days within which he will have great power—he will effectively channel the power of the land itself—but if he drinks human blood in that time he will remain a vampire forever and he (the Vampire) will also be released into the world. Seeing himself as the guardian of his kingdom and its people, with his life intimately connected to both, Vlad decides to give his life in their service and absorb the energy of the kingdom to become one with the environment itself. Vlad drinks the Vampire's blood, and the power of the ecosystem around him surges through his body, transforming him into a huge swarm of bats that swoops through the nighttime woodlands below, leaving him disoriented and half-naked in the stream at the foot of the mountain the following morning.[59]

On returning to the castle, he discovers that it is under siege by the Turkish army. As the sun goes down, Vlad walks out of the castle and kills the oncoming hordes singlehandedly, harnessing the environmental energy once again to become an all-consuming tsunami of bats. After this victory, he tells his people to leave the castle for the mountain monastery. Mehmed (Dominic Cooper) regroups and sends a huge army of men to conquer the refuge, and once again Vlad goes out to meet them, but they prove to be a decoy, and a small group of men secretly enter the monastery to kidnap Ingeras. Trying to save the boy, his mother, Mirena, is pushed from the tallest tower and falls to her death—Vlad is unable to return in time to save her. With her dying words, she tells Vlad to drink her blood and go forth to save their son and the kingdom—the land and its future—knowing he will act wisely with

---

59. Elements of this connection to the landscape are seen in the figure of the queen in *Snow White and the Huntsman* where Queen Ravenna can change into a flock of ravens and influence parts of the kingdom she is creating.

the power he will be given. Vlad complies and receives control over the entire ecosystem, causing storm clouds to mass overhead so that he can move during the day; the inability to walk in sunlight and an extreme allergic reaction to silver appear to be the two main afflictions of becoming a vampire in Shore's film. He sires other vampires from his surviving people, and they join him in his journey to confront Mehmed. They reach the Sultan's camp, and while Vlad hastens to rescue his son, the other vampires attack the Ottoman troops. Mehmed uses piles of silver to weaken Vlad, visually showing the power of money and manmade wealth/consumerism and the danger it poses to the environment—the silver is also the tribute extracted from the kingdom itself, signifying that the Ottomans want to exploit the kingdom for their own gain but not for the good of the land. Vlad struggles, reduced as he is to the strength of a mortal, and the Sultan is on the verge of staking him when, with a last surge of bioenergy, the vampire turns into a host of bats to gain the advantage. In human form again, he stakes Mehmed and drinks his blood to restore his strength. Reunited with his son, Vlad emerges to find his vampire army waiting for him but realizes they will be a danger to his son and the future of the kingdom and so causes the storm clouds above to separate, allowing the sunlight to burn them to ash, including himself.[60]

In many ways, this signals the true ending of the story. It shows Vlad, not unlike Stoker's Dracula, as an entity that is at one with his home environment and able to both control it, as seen in the weather and storm clouds, and take on the form of its fauna against those who threaten his domain. However, he also realizes this is a dangerous power that makes him monstrous to his own kind while also positioning him as the center of the ecosystem's balance and power. Now that the threat of the Ottomans has been destroyed, the "monster" that Vlad has become is no longer needed, and so as steward of the ecosystem, he willingly reintegrates himself with the land. The vampire's connection to all parts of the environment is seen even more clearly in the next movie, which replaces a human monster with a more exotic one.

Primal returns to the present day but has its beginnings very much in the past. Like a few films mentioned previously, the "vampire" in this film is as old as the land it is joined to, and it seemingly exists quite happily as long as it is left undisturbed. This marks *Primal* out as different from *Burnt Offerings* and *Jug Face*, where the creature actively seeks sustenance by luring the unwary into its deadly grasp. The story is set in Australia, where a group of friends set out to look at some ancient cave paintings in the outback—there is much of Australian Gothic here with the almost supernatural dangers of pre–White-colonial landscapes consuming the present (see Prosser 2018, 87–96).

---

60. Of course, the film's producers, Universal Studios, had other ideas and resurrected Vlad to continue the story into potential sequels.

The long car ride they undertake parallels Harker's journey to Transylvania, signaling a journey back in time but also a journey to a terrain where humans are not welcome. Some of the post-colonial aspects of the narrative are negated by the opening scene of the original aboriginal cave painters suffering the same fate that awaits the white youth of 21st-century Australia. They finally arrive at their destination but discover that the quickest way to the site is through a very claustrophobic tunnel to the other side of a rocky outcrop. One of their group, Anja (Zoë Gumeau), cannot cope with tight spaces and starts to panic, cutting herself on the jagged interior of the passage as she loses consciousness. The group backtrack out of the tunnel and tell Anja to follow the road to the site on the other side of the outcrop, a journey that will take a few hours, while they go back through the shortcut. However, they have failed to notice that the blood from the cut has been greedily absorbed by the tunnel floor and awoken the creature that lives in its walls.

The group emerges from the passage into glorious sunlight and an amazingly fecund landscape of lush plants and undergrowth. Anja eventually arrives with the car, and they set up camp and explore, finding rock paintings that show strange symbols and a monstrous figure drawn near a shape resembling a vulva or an aperture. Suddenly, Anja is attacked by some kind of mutant rabbit with razor-sharp teeth—vaguely reminiscent of the creature that attacked the attendant in *Splinter*—that manages to bite her. Her friends manage in turn to kill the creature but are clearly unnerved by it. They return to their camp, where the insects, particularly the mosquitoes, appear unusually aggressive. That same evening, one of the members of the group, Mel (Krew Boylan), decides to go skinny-dipping with her boyfriend in a nearby pond. He declines her invitation, but she proceeds on her own. She emerges covered in huge leeches. The rest of the group rush to the scene, and although they manage to remove all of the creatures, it is not long before Mel starts feeling unwell. During the night she breaks into a high fever, and eventually her teeth start falling out. The group decide to wait until morning to reassess her condition before seeking help—there is no phone reception to call the outside world.[61] However, when morning arrives, Mel now has rows of razor-sharp teeth and superhuman strength. She proceeds to attack her friends, finally grappling one of them, Warren (Damien Freeleagus), to the ground and ripping his throat out.

Mel takes Warren's body, drags it to the mouth of the tunnel and throws it inside. Another member of the group, Dace (Will Traval), appears ill after swimming in the water, though the source of his illness does not appear to be the leeches, and there are signs that it is actually the environment itself

---

61. The loss of phone signal/WiFi often symbolizes the movement from civilization/the present to an environment more savage/from the past.

Vampiric evolution. Krew Boylan as Mel in *Primal*. Directed by John Reed (London: Kaleidoscope Entertainment, 2010).

that is attacking them. Anja finds a plastic water bottle near the pond that has been partially eaten, not by an animal but by the swarms of tiny flies that hover near the water's edge. The group decides to try and drive toward help but discovers that all the vehicle's tires have been dissolved by the muddy soil beneath them, as though the whole environment has been harnessed to attack them and prevent their escape. Dace seems to be suffering in the same way Mel did before she went mad, so the three survivors decide to kill him, but before they manage to do so he mutates and attacks them. Anja and Chad (Lindsay Farris) escape, but Dace chases Kris (Rebekah Foord) until they run into Mel, and the two mutated monsters throw their former friend into the tunnel, though the creatures themselves will not enter it. Anja witnesses this scene and tells Chad that the passage is their only way of escape, but when they almost reach its entrance, Chad attacks Dace and both are killed, allowing Anja to enter the tunnel.

The tunnel prefigures the primal cave in *Dracula Untold*, not least in the way it acts as the focal point, or dark heart, of the environment beyond, and it can likewise morph in tandem with the environment—a more extreme version of *Jug Face*. Anja controls her claustrophobia and struggles through the tunnel until she suddenly comes upon Kris in a wider section of it. Kris is now in the advanced stages of pregnancy and has obviously been impregnated by whatever lives within the cave. As Anja stops in her tracks with shock, Kris grabs her friend's blade and slices open her own stomach to kill the creature inside of her. Before Anja can move, snake-like tendrils appear from the walls and wrap around her wrists and ankles, pulling her to the floor. The creature appears from a hole/crack in the floor, a fluid entity that rears up like a large penis ready to impregnate her. Anja manages to break free from her bonds and snatches up the blade, slicing into the entity and enabling her to escape. The entity may have been killed by the blow, but as with Stoker's vampire and its later kin, this seems unlikely.

The film suggests that the creature has always existed in the tunnel—there is no suggestion in the cave paintings of a time before it took up residence there—definitively linking it to the localized ecosystem and to a time predating the notion of a Christian God. Although there is no overt use of religious symbolism in the narrative, the framing of the creature as predating the arrival of European colonizers suggests location-specific, god-like entities. In this sense, the creature is not unlike that in *The Ritual* (to be discussed later), *The Lair of the White Worm*, and *Jug Face*, each of which feature entities old enough to be god-like, though unlike them, *Primal*'s vampire does not seem to require a worshiping congregation or a "Renfield" to assist it or feed off of. It is consequently a very localized environmental regulator, existing neutrally within the rocks around the tunnel and only aroused when humans enter its domain and shed blood.

It is unclear if, as in *Burnt Offerings*, the energy released by the victim's blood is used to provide extra sustenance for the wider ecosystem over which it holds influence. The need for the creature to reproduce itself likewise remains unexplained, whether as a means to spread its influence over a bigger area, consolidate control, or create "heirs" to perpetuate its line. However, humans are of particular importance to its survival—there appears to be very little diversity in the fauna in the area and no long-term predators larger than leeches. This creates a curious ambivalence toward humans entering its domain, as the creature seems to kill every human that comes near it yet also requires females to express its suppressed organic fecundity. In a sense, the environment benefits from its distance from civilization, but not all such ecosystems can remain so, and the final two movies discussed in this chapter look at areas of direct human incursion and the vampiric response they produce.

The Witch, Robert Eggers, 2015 / *The Hallow*, Corin Hardy, 2015

Robert Eggers' film, like many discussed herein, seems to be neither about vampires nor about ecology, set as it is in a pre–Salem witch trials New England in the early 17th century. However, it strongly engages with the theme of unwelcome colonization of new lands—not unlike *Primal*—and the use of vampiric fauna and flora to discourage human incursion. The story itself appears to foreground the figures of witches and the Devil, though as I have written elsewhere (see Bacon 2019), there is much correlation to Stoker's *Dracula* in the film in terms of specific character roles and vampire tropes—particularly around bodily transformation, familiars, and "brides"—but the alien landscape and religious extremism that underpin the narrative allow for other interpretations as well.

The film begins with a family on trial in a Puritan settlement in New England in the 1630s. They are newly arrived from England, and the community has a large protective fence around it, which seems aimed as much at keeping the community in as it is at keeping the dark forces of the New World out. Indeed, the film is set at a time when Satan was believed to be a real and present danger walking the land in search of souls for his taking (see Millar 2017, Hall 2005, and Demos 2004). As such, the environment outside the compound is one that is inherently dangerous, a world beyond the strictures of spiritual and social guidance that keep the faithful safe. The family that is about to be expelled from this haven of the modern world is that of William, his wife Katherine, and their children, Thomasin, Caleb, Mercy, Jonas, and baby Samuel, as punishment for William (Ralph Ineson) having his own ideas about religion and thinking the community members are ungodly. The family are summarily forced out into the uncivilized, ungodly wilderness with the few animals they have, including a large ram called Black Philip.[62] They set up a farmstead near the edge of a forest, but it is not long before things start going awry.

As with some of the previous examples in this chapter, ancient woodlands often harbor dark forces that do not take kindly to human incursion, particularly when the invaders are unwelcome human colonizers representing change and the exploitation of natural resources. It is no surprise, then, that other exiles from this Puritan settlement and others have found ways to integrate with their surroundings in a way that keeps them in tune with this new "old" environment and the more organic energies that vitalize it. The forest is home to witches who utilize earth magic at the behest of the native ecology. One such witch lives near William's farmstead, and while Thomasin (Anya Taylor-Joy) is looking after Samuel (Axton Henry Dube), the witch snatches the baby when the girl is not looking. The witch kills the infant for its blood in order to make herself young again—though as seen in *Primal*, this can equally energize the environment, which in turn passes its life on to her.

Like other vampiric beings in tune with their surroundings, the witch is able to take the form of the fauna around her. At various times we see her in the shape of a rabbit, a crow or flying in the night sky like an owl. The disappearance of baby Samuel causes much commotion and resentment back at the farm, where Katherine (Kate Dickie) blames her daughter for the dis-

---

62. Goats were often taken with settlers going to the Americas as farm stock.

appearance. Unsurprisingly, the family's fortunes continue their downward spiral as the crops they have sown are failing, with the heads of corn covered in mold. The state of the crops can be seen to be directly caused by the woodlands as well in an extension of the forest's intervention through the witch. The woodlands leave poor soil once cleared, but they are also a source of dark influence that prompts the growth of the mold—not unlike the bizarre, energized fungus in *Splinter*. Furthermore, the kind of mold that grows on maize has hallucinogenic properties.[63] Much of what follows in the film can be attributed to these hallucinatory effects, not unlike what occurs in the Ecohorror film *The Happening* (Shayamalan 2008) in which the Earth's vegetation produces a poisonous gas that kills humans. In this sense, the forest is purposely destabilizing the family so that it can prey on them one after the other.

Next on the list of victims is Caleb (Harvey Scrimshaw), who on an excursion to set traps with his father unknowingly encounters the witch in the form of a rabbit. In a later scene, Caleb asks Thomasin to join him in checking if anything has been caught in the traps. While in the woods, the two become separated, and Caleb stumbles across a small cottage to which the witch had taken Samuel earlier. A very beautiful woman steps out of the cottage, beckoning the young boy toward her. Caleb advances and she embraces him, her haggard hand appearing from her cloak as she holds him tightly to herself. Thomasin returns home without him, increasing the suspicion of her family that she is evil in some way, a feeling that is further exacerbated by the twins, Mercy (Ellie Grainger) and Jonas (Lucas Dawson), who call her a witch.

The twins were in part encouraged to do this by Black Philip, whom they spend a large amount of time playing with. It transpires that he speaks to them, and while it is later attributed to his being Satan, he can also be seen as the dark influence of the woodlands. Indeed, it is suggested that the goat has been with the family for a long time and probably made the trip over to the New World with them, but it is only now that he is in the vicinity of the dark energies from the woodlands that he begins behaving with directed menace.

The malevolence of Black Philip as a manifestation of the spirit of the forest begins subtly with the youngest children but becomes increasingly interventional and physical as the film gets nearer to its conclusion. Just as the family are trying to cope with Caleb vanishing, the boy suddenly returns, appearing naked outside the farmhouse, where Thomasin finds him. She takes him inside, where he is bedridden, and it is not long before he is writhing and delirious, seemingly reciting a prayer of some kind. Mercy and Jonas suddenly begin chanting as well, and all eyes turn to Thomasin as if she is the source of the evil overtaking the family. Caleb begins reaching upward

---

63. Anon, "The Witches Curse: Clues and Evidence," *Secrets of the Dead*, 4 June 2014, https://www.pbs.org/wnet/secrets/witches-curse-clues-evidence/1501/, accessed 9 June 2019.

and talking as if he is about to meet God and then slumps to the ground and dies. Thomasin runs from the room and William follows, accusing her of evil deeds, but she says it is the twins who have made a pact with Satan in the form of Black Philip. Her father, on the verge of hysteria himself, locks all the children in the barn with the goats. It is at this point that the evil influence of the forest overwhelms the farmstead, and the hallucinations reach a crescendo; the twins awaken to see a naked woman drinking the blood of one of the goats, who turns toward them and cackles; Katherine thinks she sees Caleb alive and sitting in the corner of her bedroom cradling baby Samuel in his arms. She goes to him and takes the baby so that it can suckle from her, but it is revealed to be a large crow.

The following morning, the dreamlike quality of the scenes continues, and William awakens to find the barn partially destroyed with no sign of the twins. Thomasin emerges from the part of the barn that is still intact to see Black Philip charging at her father and impaling him with one of his horns. The black ram pushes William into a pile of logs that subsequently collapses on the man, killing him. At this point, Katherine steps out of the house and sees her husband dead and no sign of her children. Thomasin appears to be the only one alive and, in Katherine's mind, must be responsible for what has happened, so Katherine attacks her daughter. With her mother's hands around her throat, Thomasin reaches for a knife and, while repeating "I love you" over and over, cuts open her mother's throat. The forest has almost completed its work now. Using Black Philip and the dreamlike state it has induced in the family, it has killed nearly all the family so that it can consume/absorb their energy into itself (not unlike the grounds in *Burnt Offerings*)—the twins were taken by the witch to be sacrificed, and Katherine and William are both bleeding out on the ground. As a result, the spirit of the forest has also separated Thomasin from her family and the society she was once a part of, making her ready to become one with the primal environment around her.

This last is worth explaining further, as Thomasin has only ever wanted to be good and to be looked after. She wanted to stay in England but was forced to leave; she wanted to stay in the settlement but was forced to leave; she wanted to be loved by her family, but they forced her to leave—Katherine told William to sell her to another family. Through all of this, she only wanted to be accepted by God, but he never gave her a sign or an answer. Consequently, she decides to turn to Black Philip and the dark energies he seems to embody. She walks to the barn and offers herself to him if he speaks to her, and unlike God who ignores her, she hears the goat talk.[64] Not only that, he takes a human form to offer her a contract to sign so that she may "live deliciously" thereafter. This suggests a parallel between Philip and the witches,

---

64. The human shape and voice of Black Philip is provided by Daniel Malik.

and indeed the vampires in the films discussed earlier, who can change form to influence and lure their prey into a trap. In a sense, they become "Renfields" to the ancient woodlands, assistants that bring not only food but also new supplicants; this relates to films like *The Ritual* (mentioned later) that need not only sources of food but also willing followers.

By now it is nighttime, and Philip guides Thomasin out of the barn and into the woods, telling her to take off her clothes and walk toward a clearing in which a fire is burning—a symbolic shedding of the world she is now leaving to be provided with new clothes that belong to this ancient ecosystem. Naked witches surround the fire, slowly rising into the air and embracing the darkness of the night and the energy of the forest. Thomasin spreads her arms and begins to rise, abandoning the modernity of the Old World for the ancient powers of the New and an environment that protects itself from outsiders. The story ends there, though the setting of it 60 years before the infamous witch trials in the area suggests that the dark forces of nature will never truly die. The final film in this chapter equally revolves around an ancient forest and the idea of an infectious, mind-altering fungus but in a far more modern setting.

The Hallow is, arguably more than the other examples above, about eco-colonialism. Whereas *Vampire Bats* shows the wider ramifications of human incursion into natural habitats, Hardy's film speaks of the continuing exploitation of a land and its ecology by an imperial power, i.e., Ireland by the English.[65] It begins with a quote/poem: "Hallow be their name, And blessed be their claim. If you who trespass put down roots, Then Hallow be your name. (*The Book of Invasion* C.1150)." *The Book of Invasion* (Lebor Gabála Érenn, also known as *The Book of the Taking of Ireland*) is an 11th-century collection of poems and prose that mythologized the preceding history of Ireland, telling of six tribes/peoples that have "taken" the land, ending with the Gaels (the Irish people). The film then shows an English conservationist, Adam (Joseph Mawle), traveling to a remote part of Ireland with his wife, Claire (Bojan Novakovic), and baby Finn. They arrive at a cottage on the edge of a large and ancient forest where they will be staying while Adam, a plant and fungus specialist, works out plans for a logging company to begin its work. Not long after their arrival, Adam ventures into the forest with Finn strapped to his back to begin looking about and assessing the condition of the trees. They come across a small, abandoned house not far from their own, and Adam ventures inside to discover a dead animal in a corner. Upon closer

---

65. Eco-colonialism can be used in two ways, either meaning the exploitation of a country's ecology/resources by a colonizing nation or the forms of colonization enacted by conservation and ecology programs/bodies, particularly those promoting tourism to boost funds to run conservation programs but which adversely affect traditional culture and heritage sites, etc.

inspection, some kind of fungal growth appears to have burst out of the body and caused the animal's death. Adam decides to take a sample of the fungus to examine back at the house and carefully scrapes a lump of the black, glutinous substance into a test jar. While Adam is away, Claire is visited by their nearest neighbor, Colm (Michael McElhatton), who is angry that Adam is away and has not heeded his warnings to remain indoors. Adam returns and later that night places a sample of the fungus on a slide to examine it. What he sees is strangely reminiscent of the polyp in *Nosferatu*, a cell with long feeler/tentacles encasing and "biting" a nearby healthy cell as though it is vampirizing it.

While Adam works, Claire checks on Finn in his cot upstairs, only to find his quilt covered in a black, gooey substance. She calls Adam to investigate the attic, assuming a leak of some kind is the source of the gooey substance, but when he looks, he can find nothing, failing to notice that strands of the fungal growth are making their way across the beams. Claire and Adam head downstairs for dinner, but the window in the baby's room shatters, prompting them to rush back upstairs. They can find nothing but call the police nonetheless, thinking an intruder might have broken into their house. When the policeman arrives, he looks around but can find nothing, warning them of the local legend about The Hallow, the faerie people that supposedly inhabit the forest and kidnap babies, though he professes not to believe the tale himself. The policeman leaves, and Adam, unable to sleep, heads outside with their dog to photograph evidence with a flash camera. He hears odd noises, and when he takes a photo, pointing the camera toward the trees, the flash of the camera seems to reflect in a pair of eyes. The dog is clearly riled and growls, but they return to the house.

The fungus in *The Hallow* closely resembles the fungus that oozes out of the tree at the end of *Splinter*. In this film, too, the fungus is shown bleeding out of trees in the forest, almost like a concentrated essence of the forest itself, alive and sentient. The following day, Adam takes Finn with him on a drive to get the window panel fixed in the nearest town. There, he is again warned about the creatures that live in the woodlands. Colm visits their house again and leaves an old book for Claire to read that tells the story of The Hallow. On his way home, as Adam reaches the edges of the forest, he loses control of the car and nearly crashes it. He leaves the car and examines the engine, finding it covered in tendrils of the fungus. He then circles around to the trunk in order to find a cloth with which to clean the fungus, but as he leans in, he is knocked unconscious. When he finally wakes up, it is dark, and he hears Finn screaming. He realizes he is locked in the trunk of his own car and manages to break through the back seat, grab Finn, and escape onto the road. Looking back, he sees the side of the car has huge, jagged scratches down the side. He runs back to the house and tells Claire to go upstairs and ring the police. He gets his gun, but when he returns downstairs, the house has been ransacked. Thinking his neighbor responsible, Adam instructs Claire to pack, and they all head to the car to leave. The fungus has now covered the entire car engine, but they clear it off and start the vehicle. Creatures have appeared from the forest by this point, seemingly growing out of the woodlands as a human-shaped embodiment of it, and start throwing lumps of the fungus at them as the family drives away, causing them to crash. They escape back to the house and lock the doors, turning on the lights, as light seems to repel the creatures, not unlike certain vampires whom sunlight can kill. Adam looks out of the keyhole in the front door to spot the creatures, but one of them injects fungus into his eye, causing him to collapse on the floor. The lights go out, and Adam heads downstairs to start the generator while Claire runs to the attic with Finn and locks the door behind her. One of the creatures chases after her and infects the wood of the loft door with fungus, allowing its arm to pass through it—as with other vampiric entities, if something is "infected" by its "bite," it becomes part of the vampiric body. As it extends its arm toward Claire, its fingers seem to blossom into flower-like stamens that try to drip fluid fungus into her eye, but Adam manages to start the generator and the lights flash back on, scaring the creatures away.

Meanwhile, one of the creatures grows through the wooden floor downstairs and snatches Finn. Adam loses consciousness in the course of attempting to chase them, but Claire manages to save the child and return to the house. Adam has read parts of the book in the meantime and realizes that The Hallow swap babies for changelings and take the human ones for their own. By this time, Adam is hallucinating due to the effects of the fungus infecting his system—not unlike the effects of the mold in *The Witch*—but through this he knows that Finn is a changeling and must be killed. Claire stabs Adam and runs off with the baby. Adam is now transforming into one of the creatures, with the infection spreading from his eye throughout his body. He makes his way to the creatures' "nest," which seems to be under the abandoned house, and discovers the real Finn—not unlike the Aswang, most of their number appear to be made up of former humans who have become one with the forest and gain a level of immortality by doing so. Adam grabs the baby and, with his last vestiges of humanity, runs to find Claire. She is surrounded by the creatures and only keeping them at bay with the camera flash. Adam beats back the other creatures and puts the real Finn on the ground, telling Claire that she is holding the changeling. She finally believes him and picks up the baby Adam has brought just as her husband is fatally wounded by one of the creatures. By this time, dawn has broken out, and beams of sunlight are beginning to stream through the tree canopy, allowing Claire to make it safely out into the daylight. Back in the forest, the sun begins to shine on the baby left on the ground, and momentarily Adam thinks he had given his wife the wrong one, but suddenly steam starts to rise from its body, and the flesh on its face begins to blossom like a large flower, revealing its ashen skeleton beneath—a tiny creature of desiccated wood. Claire and Finn escape, but the story is not over; while the end credits of the film roll, we see a logging company—presumably the one that Adam worked for—chopping down a large number of trees and loading them onto waiting lorries. As the camera pans across this scene, it focuses on the logs stacked on one of the trucks that is preparing to depart, revealing on the underside of one of them copious amounts of fungus oozing out of the wood. Just then, the vehicle pulls away, heading toward the outside world and civilization.

This last film presents a vampiric entity in the vein of *Splinter* with the life-force of the ancient woodlands going out into the world to wreak revenge and discourage humankind from entering its domain—indeed enacting a kind of reverse colonization similar to that of Stoker's *Dracula* (see Arata 1990). As mentioned at the beginning of this chapter, all these examples demonstrate ways in which ecosystems attempt to protect themselves and restore balance by utilizing vampires or vampiric elements within them to either lure victims and kill them or create active agents, enacting a battle between the past and the present to recreate a time when humanity had a more respectful and symbiotic relationship with its environment. The next chapter follows this thread, continuing the logic seen in *Splinter* and *The Hallow* to the point where the ecosystem creates a vampire apocalypse to rid itself of the human plague that is intent on destroying it.

CHAPTER 3

Undead Eco-Warrior
The End of the World as We Know It

This chapter looks at the apocalypse, specifically those moments when the planet unleashes a vampiric plague aimed at humans to reset the ecosystem.

Tales of dystopian futures and the end of mankind, other than biblical depictions, began to gain popularity in Western literature in the 19th century, seemingly a time that engendered such thoughts in the popular consciousness. Mary Shelley's *The Last Man* (1826) is an early example, but by century's end, writers such as Richard Jefferies (*After London* [1885]) and H.G. Wells (*The Time Machine* [1895] and *The War of the Worlds* [1897]) encapsulated something of the anxieties around the end of the empire and the new century. After the two world wars of the 20th century and the creation and explosion of nuclear weapons, such post-apocalyptic narratives took on a new urgency, as their visions of a cataclysmic future were all too close and real. A seminal text in this regard, and more importantly in manifesting the apocalypse's uniquely vampiric character, is Richard Matheson's novel *I Am Legend* from 1954. While this is obviously not a film, though it has been used as the inspiration for many, it is worth beginning this chapter with a closer look at how the story manifests an ecosystem fighting back against the human threat to its future.

I Am Legend, Richard Matheson, 1954 / *The Passage*, Justin Cronin, 2010/ Liz Heldens, 2019–present

Matheson's novel is set in the 1970s, and even though it suggests an ongoing nuclear conflict producing dust clouds and contamination, it is not a vision of the future but rather a parallel world where World War II does not appear to have ended until nature enters the fray. In the novel, the war has produced large dust storms that blew through major cities, causing an explosion in the mosquito population. These two developments carry a form of infection, which then causes humans (and some animals) to die and come back to life as vampire/zombie-like creatures. This very much ties in to the popular idea of radiation having supernatural-like effects and the carelessness (hubris) of mankind in its application of science and (dis)regard for the environment—indeed, when the first nuclear bombs were exploded it was believed that they might cause all the oxygen in the Earth's atmosphere to ignite with catastrophic effects.[66] The resulting disease is caused by the disintegrated matter of civilization, seen in the dust storms, and the mosquitos wreak an ecological revenge on the humans for the damage they have caused. The infection thereby mirrors the desiccation of the environment by enacting the same upon the human race itself. This begins to see the resultant plague as being environmental in its origins and much more of an ecological regulator than just out-of-control science.

Unlike previous incarnations of the vampire, Matheson's vampires are lumbering creatures more akin to the slave-like, mindless Haitian zombies that were used in stories and films before World War II, though they still exhibit an extreme reaction to garlic, crosses, and mirrors—the first of these is explained as a biological mutation/allergic reaction in the novel, but the last two are seen as a kind of extreme cultural conditioning, which will be returned to later. At the opening of the book, all these originating factors seem to have quelled, and it is now only the vampires that are left to spread the disease and infect any remaining humans, of whom there appears to be only one: Robert Neville. Neville does not know why he is the only human immune to the disease but surmises that his immunity might be linked to a period in his past when he was stationed in Panama during the war and was bitten by a vampire bat. Neville himself explains it as follows:

> I was bitten by a vampire bat. And … my theory is that the bat had previously encountered a true vampire and acquired the vampiris germ. The germ caused the bat to seek human rather than animal blood. But, by the time the germ had passed into my system, it had been weakened in some way by the bat's system. It made me terribly ill, of course, but it didn't kill me, and as a result, my body built up an immunity to it [Matheson 2007, 79].

He further notes that this might have been the only bat infected and that he might have been its sole victim before he killed it. There are obviously some

---

66. "Shortly before the first detonation of a nuclear device, the Trinity test, July 16, 1945, Enrico Fermi jokingly took bets on 'whether the atmosphere will be set on fire,' according to physicist Hans Bethe. Fermi was mocking concerns raised by Edward Teller that the fission explosion might trigger runaway fusion." John Horgan, "Bethe, Teller, Trinity and the End of Earth," *Scientific American*, 4 August 2015, https://blogs.scientificamerican.com/cross-check/bethe-teller-trinity-and-the-end-of-earth/, accessed 15 June 2019.

odd bits here, not least Neville's mention of a "true vampire," which he tries to define at various points in the story as natural creatures that have been poorly explained, or misinterpreted, by superstition and misunderstanding. More importantly, it cites the original disease, or carrier of it, as a natural creature, an evolutionary vampire that lives off human blood and was bitten by a vampire bat, which then contracted the infection.[67] This version of the contagion carried by the bat is a mutation but a natural one, as though nature were preparing for such an outbreak. In this sense, there is more than a hint of Wells' *War of the Worlds* in this premise, where humans are saved by the common cold, a disease to which they are resistant because of their connection to the Earth. In Matheson's story, humankind has distanced itself from the land, no longer part of Earth's ecology but an alien, foreign body that needs to be destroyed, much like Wells' Martians. It is Neville, through no intention of his own, who remains connected to the Earth and so marked for survival, or so he thinks.

The vampires bring about the breakdown of civilization, reducing it to a simple fight for survival. Neville, whom Amy J. Ransom rightly calls "a murderous villain" (Ransom 2018, 37), conversely feels it is his ordained mission to destroy all the vampires and try to restore the world to its former state. He thereby gradually distances himself from the emerging world of ecological balance one vampire at a time. Before considering the emergent new world, it is worth examining the vampires as a means to an ecological end. As mentioned before, the vampires suffer from severe reactions to garlic, religious symbols, and mirrors. Their allergy to garlic is attributed to a reaction in the blood of the infected—this also includes a deadly response to sunlight that destroys the virus and subsequently its host—but the other two vampiric symptoms, though equally folkloric in origin, are attributed to what Neville explains as "hysterical blindness." This suggests a cultural trauma, first related to one's religious upbringing/background, causing an aversion to the symbolism of that faith, as Neville comments: "as far as the cross goes—well, neither a Jew nor a Hindu nor a Mohammedan nor an atheist, for that matter, would fear the cross" (Matheson 2007, 74), suggesting that if Christians fear the cross, then Jews would have a similar reaction to the Star of David, but Matheson's narrative does not substantiate such a view. The aversion to mirrors continues in a similar vein, with Neville surmising—in contrast to lore where vampires have no reflection—that vampires are so "mentally" loathed by the living that this would "set up a block in their weakened minds causing them be blind to their own abhorred image" (Matheson 2007, 63). These explanations are Matheson's attempts to rationalize the vagaries of the

---

67. The metaphorical "dark" lands south of the U.S. often seem to provide the source of vampiric outbreaks. See *Nightwing*, *From Dusk till Dawn*, and *The Passage* to name but a few.

supernatural, folkloric attributes of the vampire, but both serve to construct his undead as creatures/creations of the culture that produced the war and, as such, come to symbolize it even more than Neville himself does. The vampires thereby reveal the lumbering, all-consuming nature of the society that consumed and destroyed the world with no thought but of their own primal drives; nature has created a mirror of humanity, a doppelgänger that will haunt and destroy it.

In this sense, a future world that is once again in balance requires that the vampires and Neville are disposed of, as both represent a traumatic remembrance of the past, the "legend(s)" of the title. This job falls to the originally unseen third species in the story that is not named but is half vampire and half human. Members of this third species do not have the same level of immunity as Neville, but although infected, they have not been killed and resurrected by the virus but rather have managed to produce a form of medication that holds them in a kind of human-like stasis. Unlike Neville, they have no wish to return to the past but instead wish to forge a new future for themselves. Although this future is never explicitly represented as ecological, they are shown actively killing the vampires, and eventually the last human as well. At the very least this configures them as a staging post on the way to the creation of a re-established balance between "humanity" and its environment, where the previous means of domination and exploitation—the city, infrastructure, and society—are swept away to be replaced by an alternative as yet unknown.

Matheson's story has been the inspiration for many vampire/zombie apocalypse texts, and George R. Romero has said as much about his seminal zombie film *Night of the Living Dead* (1968),[68] in which the undead bear close resemblances to the lumbering hulks of *I Am Legend*.[69] More recent adaptations have tended toward the destruction of humanity, though many keep the connection to ecological rebalance. The film *28 Days Later* (Boyle 2002) shows animal experimentation as the cause of the original outbreak—the "Rage Virus" that causes the vampiric plague to escape from a government testing facility and heavily suggests that it is a scientific mutation that brings about the destruction of the society that facilitated said testing—though the vampires here are completely opposite to Matheson's. Rather than being slow and lumbering, they are driven by what the film calls "rage," which constructs them as extremely agile, fast, and violent. They have no regard for vampiric folklore or religion, and it only seems to be their unquenchable desire to bite and spread their infection that marks them as vampires. The infection seems

---

68. See Bishop 2017, 101.
69. However, in *Night of the Living Dead* the infection is blamed on a NASA probe that had been sent to Venus, picked up a strange radiation and was exploded in the upper atmosphere on its return.

to have no cure and can only be destroyed or contained, neither of which seems to be successful, as evinced by the sequel, *28 Weeks Later* (Fresnadillo 2007). In this sequel, society and civilization are once again destroyed, thereby allowing the environment to heal itself once the human plague has passed. A similar pattern can be discerned in Justin Cronin's *The Passage* (2010), which is both a return to and a development of Matheson's story.

The Passage is currently both novel and television series, though the adaptation to the small screen has so far dispensed with much of the backstory that made the book resonate with earlier versions of the trope of the approaching vampire apocalypse. Consequently, this study will focus largely on Cronin's original text. As with Matheson's novel, the story is set in the near future—relative to the novel's publication date—2016 in this case, though as with *I Am Legend*, the setting is a parallel world much like our own, but not quite. In Cronin's novel, a team of professors and students is funded by the United States Army Medical Research Institute of Infectious Diseases (US-AMRIID) to undertake some field research in the Bolivian jungle. Once there, they are attacked by a swarm of bats—not unlike those seen in *Nightwing*—that leaves many of the team wounded. Those who were bitten soon become ill, developing hemorrhagic fever, but those who survive seem to have a boosted immune system and enhanced strength and agility, if also a worrying penchant for human blood. The researchers (unwitting test subjects), as part of the top-secret test program Project Noah, are transported back to the U.S., where the government plans to use them to produce living weapons by refining the virus that causes these unusual effects on humans. To this aim, the government performs a series of experiments using undesirables from the prison system and creates 12 test subjects, including one survivor (Tim Fanning) of the original two who were brought home from the original research trip to Bolivia. The infected become largely sexless—not unlike the vampires in *The Strain*—and completely hairless with marble-like bodies revealing the veins beneath their skin, rows of sharp teeth, and an insatiable desire to drink/consume human blood. The only problem, from the point of view of the government, with the first batch of test subjects is that they are uncontrollable and deadly to anyone they can get their hands on. The lead scientist in the compound decides that the virus cannot combine properly with the subjects' immune systems and speculates that it might form a more symbiotic connection with a younger test subject, prompting the government to bring in a six-year-old orphan named Amy Bellafonte. The scientists inject her with a more refined version of the serum, and she seems to have none of the immediate side effects that the other subjects suffered from. Before she can be studied further, the 12 subjects take psychic control of their guards—Fanning, known as "Zero," has the most pronounced ability—and escape the compound, sparking the vampire apocalypse. The story then skips approximately

93 years into the future, where civilization has crumbled and colonies of humans try to defend themselves from the vampire hoards. Amy, now looking about 15 years old, has developed healing abilities and can psychically influence the 12—the test subject vampires who have each set up their own respective colonies with vampires they have sired and to whom they are psychically connected. Amy then begins her quest to kill the original vampires and allow humanity to finally rebuild itself.

The Passage resonates with *I Am Legend* insofar as the vampire apocalypse is brought about by a scientific mutation of the human body and the future of humanity is decided by vampire/human hybrids. One may also discern echoes of the ecological revenge examined in the previous chapter, where an environment reacts violently to uninvited interlopers entering its domain. The researchers, representing a particular kind of human civilization that seeks to exploit the natural world for its own gain, enact what Johan Höglund calls "American Imperial Gothic" (Höglund 2014, 157–61). This imperialism is manifested in the military personnel that take part in the research expedition to ensure the project receives ample funding, but it also changes the nature of the trip from one centered on study and understanding to that of governmental exploitation and control. Consequently, the team enters the Bolivian jungle with little regard for what unseen borders they might be trespassing or the delicate ecological balance they might be disturbing. The attacking bats, "a huge swarm that blotted out the stars" (Cronin 2010, 24), manifest the jungle environment itself: thick and lush and all-consuming. Those bitten or scratched by the bats are infected with the raw energy/fecundity of their surroundings, and the virus causes the infected body to exceed its normal limits, causing "bleeding from the mouth and nose, the skin and eyes rosy with burst capillaries, the fever shooting skyward, fluid filling the lungs, coma" (Cronin 2010, 23). While not all the diseased can survive such a biological onslaught, those who do are created as new, energized, superhuman creatures.

The effects of this biological plague separate Cronin's vampires from Matheson's. Those in *I Am Legend* seem to signify an entropy of life, an ecological vampire draining the energy from humanity, but in *The Passage*, the opposite seems to occur, where the vampires are supercharged, causing human civilization to explode with the mutants' immense strength, speed, and hugely extended lives—they live 1,000 years in the novels. As with *I Am Legend*, the global ecology does not necessarily want any species to become extinct but instead wants humans to exist as part of a symbiotic, interdependent whole. Thus, Amy's place within this environmental evolution—as she has fully integrated the energizing virus into her own body—symbolizes the nature of the new world, which also requires her to rid it of the other vampires. Again, like in Matheson's narrative, the original vampires are a means to an end, returning the Earth to an "archaic" and "primitive" form (Byron and

Stephanou 2013, 194), and once their use is over (the collapse of the human civilization that caused the original outbreak) they need to be disposed of to allow the world to reconstruct itself.

Much of *The Passage*'s post-apocalyptic landscape is shown as dry and dusty—largely due to the location of the colony that Amy arrives at and its proximity to one of the original 12 vampires—but it gives the impression that the vampires, left unchecked, drain the life out of the environment just as they do to the remaining humans. *Mad Max: Fury Road* (Miller 2015), which also features vampires/bloodsuckers, shows something of the same. This might also explain why the ecosystem will find ways to dispose of the vampires it has created, as they are as dangerous to the environment as the humans they are designed to get rid of. However, not all examples necessarily follow that route, and a more recent adaptation of Matheson's novel points to more instant ecological reclamation of the world while the vampires are surviving. *I Am Legend* (2007) by Francis Lawrence shows an extremely impatient nature that is eager to reclaim the land taken by human civilization. The apocalypse is created by scientists who believe they have found a cure for cancer, one which leaves the film's "legend," Robert Neville (Will Smith), as the last survivor in New York City. In earlier adaptations (*Last Man on Earth* [1964] and *The Omega Man* [1971]), the Neville character travels the city in search of supplies and vampires, but the metropolis has not changed in any way other than being abandoned. The skyscrapers, malls, highways, etc. are all intact and well-kept as if the inhabitants were transported out, and where signs of damage are apparent, it is minimal. Lawrence's film bucks the trend, and although the buildings are still there and only three years have passed since the original outbreak, the buildings have become almost instantly overgrown by vegetation and reclaimed by wild animals. One scene in particular highlights this point: as Neville drives down one of the main streets in the city, a herd of deer jumps out in front of his car and runs off down a side street. Neville gets out of the car and grabs his gun to chase and shoot one of the deer, but as he rounds a corner a lion pounces out, dragging the animal to the ground and beginning to eat it. Even though it is suggested that the animals have escaped from the zoo in Central Park, that does not sufficiently explain the virulence of the vegetation nor the lingering feeling that the plant life has been eagerly awaiting mankind's disappearance to reclaim its world. This idea becomes even more explicit in the next films to be looked at with their various forms of vampiric vegetation.

Voodoo Island, Reginald LeBorg, 1957 / *The Day of the Triffids*, Nick Copus, 2009

Voodoo Island, as heralded by the film's title, tries to feed off the negative press given to Haiti during the American occupation of the island pre–World

War II—many films were produced about zombies[70] and voodoo at that time, including *White Zombie* (Halperin 1932), *Revolt of the Zombies* (Halperin 1936), and *I Walked with a Zombie* (Tournier 1943), following upon the publication of William Seabrook's sensationalist memoir of his time in Haiti studying the island's voodoo practices (see Luckhurst 2015, 30–2). Yet as *Voodoo Island* unfolds its story, it becomes more and more apparent that the film is actually about primordial nature fighting back against the all-out consumerism of the modern world.

The story begins in Hawaii—rather confusingly, as Hawaii is in the Pacific, whereas Haiti and the region associated with voodoo and witchcraft is in the Caribbean—and shows wealthy hotelier Howard Carlton (Owen Cunningham) looking over a scale model of a planned resort, the Paradise Carlton. The island in question is infamous in the area as hexed and the home of mysterious voodoo practices. Indeed, the first surveying party sent by Carlton never returned, none save one, Mitchell (Glenn Dixon), who was found washed up on the beach of another island in a catatonic "zombie" state. Carlton is now putting together another team for a second excursion, this time headed by well-known television myth-buster Philip Knight (Boris Karloff), to prove that the island is safe and the perfect site for a luxury vacation resort. The team that leaves for the island consists of Carlton's representatives—Barney Finch (Murvyn Vye), his assistant, and Claire Winter (Jean Engstrom), his head designer—together with Knight, his assistant Sarah Adams (Beverly Tyler), Mitchell and his physician, Dr. Wilding (Herbert Patterson).[71] They land on the island nearest to their final destination and hire a boat for the last leg of the journey, offering large amounts of money/potential profit to the boat's captain, Matthew Gunn (Rhodes Reason), and harbor owner Martin Schuyler (Elisha Cook, Jr.), who are also persuaded to join the team. Before they depart, Mitchell escapes his doctor and dies on the dock, seemingly from fright—extreme PTSD—and the body, together with his doctor, is flown back to Hawaii. The team depart for the island but not before finding a voodoo bag containing death-curses for all six of them on the dock. They immediately run into trouble upon their arrival as a huge crab falls out of a coconut tree, almost killing one of them.

As they proceed inland, they find surveying equipment left by the previous team, and not unlike in Lawrence's *I Am Legend*, it has already been overgrown by vines and greenery—oddly prescient of scenes in the later film *Annihilation*. They look through the surveyor's transit, spot a clearing that

---

70. Haitian victims of "vodou" (voodoo) were known as zombi rather than the later spelling zombie.

71. As the opening scene cuts away, one of the plants on the model of the resort wilts and dies, dripping blood as it dies to emphasize the voodoo influence radiating out from the island.

Chapter 3. Undead Eco-Warrior

Jean Engstrom as Claire Winter falls prey to winter and the octopus plant in *Voodoo Island*. Directed by Reginald LeBorg (United Artists, 1957).

appears to have been used by the previous team and decide to investigate. The men return to the boat to collect supplies but find that their supplies have rotted and are full of maggots. The following day, Winter wanders off and stumbles across a lagoon and decides to go for a swim. As soon as she enters the water, a large plant begins to spread out its huge, tentacle-like leaves in the lagoon and quickly ensnares the unsuspecting Winter. Her screams attract the rest of the team but too late, as she is dead when they arrive.

Knight, seemingly an expert in such things, diagnoses the plant as carnivorous and that it "throwbacks to the Paleocene epoch nearly fifteen million years ago." Adams is soon attacked by another plant that has a snake-like head on top of a large stem that tries to grab hold of her, but this time the men arrive in time and hack at the plant's roots until it releases her. The party builds a large fire and hunkers down for the night, but Finch awakes to find another one of the snake-headed plants bearing down on him, and he hurriedly runs away. He arrives at a clearing where two small children are playing, and as he watches, one of them steps into a huge, flat plant, triggering its large leaves, which snap shut around the child and crush it to death. The shock sends Finch into a catatonic state, like Mitchell before him. The rest of the team are then surrounded by natives who march them all off to their camp, where their leader shows them voodoo dolls made in their image, suggesting that the energies of the island are focused through these effigies and into the person they represent. Before restraining them for the night, the leader explains that the natives coexist with the vampiric environment because they are "survivors that fled from island to island as your world moved in to crush ours. This island always has been taboo. When we first discovered the plants, we knew why no one had lived here." Most of the surviving group soften their view of their captors, but Schuyler, still swayed by his own greed, insists that he will "not be robbed of my part of this island." While the rest are asleep, he tries to escape but falls off a rope bridge to his death. As the others hasten to the scene of the commotion, his voodoo doll materializes where he fell, signifying to the others that the magic is real and that the village leader could kill them by channeling the energy from the island at any time he chooses. The survivors' new acceptance that the ancient natural powers still hold sway over the modern world satisfies the natives, who then allow the interlopers to leave.

In many ways, *Voodoo Island* shows a very different natural environment from those previously examined in this study. Although Bruce F. Kawin calls the plant life "ruthless ... hungry and relentless" (Kawin 2012, 80), it is not intent on the complete destruction of mankind but rather demands recognition as part of a symbiotic relationship that should not be exploited. The island itself is the prime example of this interdependence, where the vampiric plants will eat the unwary—i.e., those who do not respect/recognize the nature of the environment around them (and it is the elders' duty to teach this to the children)—but they equally supply a certain amount of protection, as other communities on nearby islands leave the natives well alone (something of this is seen in *Surviving Evil*, discussed in the previous chapter). Consequently, the island acts as a balanced and contained ecosystem that is proactive in protecting itself.

The idea of voodoo is integrated differently from earlier films. In *White Zombie* and *I Walked with a Zombie*, for example, voodoo is used as a means for acquiring power and individual gain and more often than not involves the possession and control of white women (always a ready-to-hand technique for inspiring panic in a white male audience). These instances of voodoo also inevitably involve the mindless enslavement of people to the voodoo practitioner's cause, again often for financial gain, and there is obvious magic/witchcraft involved with ritual/ceremony/spell casting. This is not the case in *Voodoo Island*. Although there are voodoo dolls, and an Uhuanga, or voodoo bag, is found at the dock, the film creates the impression that the great antiquity of the island itself produces the magic with no obvious spells or rituals performed. Indeed, at the end of the film, it is the ecosystem itself that appears to allow the survivors to depart, with the village elder simply informing them of the consequences should they come back or encourage others to arrive. This warning helps define the vampiric credentials of the island and its vegetation in that they embody and gain their power through great age as well as through enacting a traumatic rupture of the present by the past, feeding off of today to maintain the vitality of history and a time when the world was in greater environmental balance.

The intersection of the ancient past/vegetation and vampirism is also seen in the film *The Ruins* (Smith 2008), in which a group of vacationers believe they are being attacked by a group of natives in a remote part of Mexico. The tourists take refuge in a vine-covered Aztec temple only to later discover that the natives were not attacking them but trying to warn them off and that the vines are in fact vampiric, having previously fed on/been created by the blood sacrifices that took place in the temple hundreds of years before. The vines invasively burrow beneath the skin of their victims, but unlike the plants in *Voodoo Island*, they seem prepared to spread beyond their ancient environment out into the world. Consequently, the villagers who are aware of this do everything in their power to prevent people from entering or leaving the ruin, thereby suppressing the ecology rather than living in symbiotic balance with it.

The Ruins reflects the ways in which plant life has become increasingly proactive, not only protecting its own environment but re-stabilizing the ecosystem of the world. One of the more well-known texts in this regard is John Wyndham's *The Day of the Triffids* (1951), which utilizes post-war fears, not unlike Matheson's novel, but its vampires are huge animated plants with a taste for human blood. Wyndham's story is largely told from the perspective of Bill Masen, a biologist who works with triffids and who suspects the tall, carnivorous plants were created by the Soviet Union and released into the wild. When the narrative opens, he is in the hospital with his eyes bandaged after venom from the killer plants splashed in them, sparing him when a green meteorite shower blinds a huge proportion of the Earth's population—Masen once again suspects Soviet space satellites might have been involved. In the ensuing chaos, various factions rise and fall as Masen and his group eventually make it to the Isle of Wight off the south coast of Britain to hatch a plan to kill the triffids. There have been various film and television adaptations of the story, but of special interest here is the 2009 two-part mini-series by Nick Copus.

Copus' mini-series links quite nicely to *Voodoo Island*, as it relocates the original London-based triffids to Zaire in Africa, constructing them as part of an ancient culture that remains in balance until the intrusion of contemporary consumerist society. Early scenes in the series show a rain-drenched woman reaching frantically toward a young boy, with the intimation that her hysteria is a response to a triffid attack, followed by the appearance of a black figure holding an ornate native mask—resonating with the voodoo images used in LeBorg's film. The human intrusion into nature is perpetrated by Bill Masen's parents, who are biologists who have come to study the mysterious plants to exploit their natural properties. Bill's father, Dennis (Brian Cox), discovers that the triffids produce an oil that can replace the planet's diminishing fossil fuel supplies, not only solving the ongoing energy crisis but also reversing the damage it has caused the ecosystem. As noted by Dawn Keetley, the exploitation of the triffids is not a solution to the planet's problems but another form of biological slavery, where "[the triffids] served *human* ends, helped save *us* from a destruction we had long been bringing on ourselves" (Keetley 2017, 21).[72] Dennis then begins to bio-engineer the plants to maximize their production of oil but, unfortunately, without curbing their dependence on human blood and flesh.[73] The interdependence between humanity's need for the flesh of the triffids and the plants' requirement of human flesh explicitly reflects the symbiotic relationship between mankind and the Earth.

The story skips forward to the year 2009, where Bill (Dougray Scott) is one of the main researchers/specialists on triffids and is working at one of the many plants nationwide that grow them for their oil—in fact, the whole world is now dependent on the plants for their energy. Everything seems to be going well until a solar event blinds a huge proportion of the world's population—this mass-blindness is shown to be the fault of experts who greatly underestimated the power of the event, prompting millions to view the sky as it occurred. Even this is not as catastrophic as it might have been until a plant-rights activist intervenes. This is a curious plot device of the film, nodding to the earlier *28 Days Later*, which was heavily based in turn on Wyndham's original *The Day of the Triffids*, where the apocalyptic outbreak of flesh-eating monsters is due to animal-rights activists breaking into a government facility and unsuspectingly releasing some infected test-monkeys into the world. This cinematic allusion also gestures toward the in-vogue issues of ecology and environmental activism at the time of the film's release, or what Sasha Matthewman describes as "contemporary anxieties about genetically modified plants, global warming, biofuels and energy shortages" (Matthewman 2010, 124). Indeed, it rather muddies any message the story might be trying to put across, because while it superficially appears to promote skepticism about society's dependence on technology—and particularly a sighted society—it equally suggests that government experts should be left alone and

---

72. Consequently, having the plant's homeland in Africa directly correlates the exploitation of them to slavery and sees their "uprising" against their human captors on par with that of the zombie horde led by "Big Daddy" in *Land of the Dead* (Romero 2005).

73. The Italian giallo film *Island of the Doomed* [La isla de la muerte] (von Theumer 1967) also features bio-engineered carnivorous plants but of the more stationary variety.

that overtly trying to save certain plants/species, etc. is foolhardy and potentially life-threatening.

Returning to the film, a rogue activist is caught at the Kingston triffid compound south of London and locked up overnight at the facility. Bill was involved in the altercation to prevent the intruder causing any damage and was temporarily blinded by triffid venom—the plants have evolved, learning to flick venom at their victim's eyes to incapacitate them in order to catch them more easily—and he is rushed to a hospital in the capital. When the solar flare blinds everyone, the activist is in a room without windows and is therefore also still sighted as power failures start to affect the facility. He escapes from his room, heads to the control room and opens all the gates to the triffid compounds, releasing the male plants, which are purposely kept separate from the females to keep the population in check—if they are allowed to propagate, their numbers will become overwhelming within a matter of months. With all the barriers down and the staff blinded, it is not long before the triffids leave their compounds, kill all the humans, and escape—it is never made explicit what parts of the humans are eaten by the plants, though most of the deaths seem to involve copious amounts of blood, suggesting it is the more vampiric menu they prefer. Meanwhile, Bill has woken in his hospital bed to the sounds of the ward in chaos. After removing the bandages from his eyes, he realizes that everyone appears to be blind.

The scenes in London are extremely important to how the film is read, as the sudden blindness does not just affect the human population, as is the case in the earlier adaptations of the story—*The Day of the Triffids* (Sekely 1963) and *The Day of the Triffids* (Hannam 1981), where the buildings are rarely damaged—but is depicted as an apocalyptic event. Some of this is due to an aircraft crashing into the center of London near the hospital where Bill is staying, but the mayhem appears to extend through the entire city as though bombs have been dropped everywhere. This scene creates many historical and generic references alluding to the Blitz and World War II with images of Churchill and the use of Whitehall as a center of operations. It likewise gestures toward *28 Days Later* with its use of private militia groups vying for ascendency and further makes clear reference to Matheson's *I Am Legend* with the hordes of blind people acting just like the zombies from the first film adaption of Matheson's novel, *The Last Man on Earth* (Ragone 1964). This last reference resonates in one scene in particular in which Jo Playton (Joely Richardson), a radio broadcaster, tries to help people in the streets around Moorgate Tube Station—in the city and underground while interviewing people who managed to avoid the solar fireworks—but she is quickly surrounded by figures moaning and grabbing at her. She is soon overwhelmed in a fashion that echoes similar scenes in zombie films of a noisy person literally consumed by the crowd, but here the zombies desire eyes (sight) rather than

The triffids are revolting. *The Day of the Triffids,* **directed by Nick Copus (BBC Television, 2009).**

brains or flesh. Bill arrives in the nick of time, his hospital apparently near the tube station, and tells her to keep quiet so they can make their escape.

This reworking of *The Day of the Triffids* therefore reimagines the blind as zombies and the plants as vampires, with both sets of "monsters" working at the behest of the environment to destroy human society so that a new ecological balance can be negotiated. After much meandering, the triffids are not defeated and manage to pollinate, but Bill finds a way to deceive the plants so that they no longer hunt his group, which in some measure points to a conclusion similar to that of *Voodoo Island.*

Bill and Jo end up at Dennis Masen's house in the country—something of a nod to the traditional, rural, non-urban, non-consumerist lifestyle—where Dennis imparts much of his research to his son. It transpires that while Dennis sought to modify the plants for his own personal gain—not unlike Schuyler demanding his part of the "island" in *Voodoo Island*—Bill's mother sought to better understand the plants and made many sound recordings of the triffids' communications with each other. Inspired by this discovery and by conversations with his father, Bill begins to remember more of his final traumatic moments with his mother and begins to realize that the native mask, which is currently in his father's house, does not just represent the archaic, supernatural nature of Zairean (African) culture but is also connected somehow to the triffids. Consequently, in the denouement of the film, as the recordings of the triffids have drawn hundreds of the plants to the house, Bill remembers that the mask is especially designed to deliver a small dose of triffid venom into the wearer's eyes, which does not make them blind but rather makes the plants blind to them—it is never explained why this should be so. Bill and his companions safely escape through the vast number of triffids outside his father's house, with their acceptance of the significance of the mask providing the same kind of protection as that afforded to Knight, Adams and Gunn in *Voodoo Island*.

This posits a symbiotic bond of sorts between the natives of Zaire, or at least those living where the triffids grow, and their environment so that even their age-old rituals and religious artifacts are linked to the flora and fauna of their surroundings, effectively positioning them as one with their landscape. This coincides with LeBorg's film, where the primitive magic of voodoo is somehow linked to the carnivorous plants and the island itself, showing that even these hybrid, "human"-ized (Westernized/colonialized) versions of the original plants can still respond to the call of their ancestral home and legacy.

After Bill and his friends escape, they head for the only place they know where there were no triffid farms, the Isle of Wight off the south coast of Britain, effectively leaving the mainland to the killer plants. The humans seem to have no plan in place for how to effectively combat the plants, armed only with basic weaponry and the lure provided by the recordings of the triffids communicating with each other. Consequently, as with the brief scene in Lawrence's *I Am Legend*, the environment has reclaimed the modern world, although how the huge number of triffids will survive with humans as their only food source is never explained. Potentially, as with other examples, once the human threat has been extinguished, the "vampires" are no longer necessary and can be allowed to dwindle until the problem arises once again. The place of vampires as planetary pest-controllers in the ecological new world is examined more closely in the next two films, which present very different perspectives on the vampiric future.

Daybreakers, The Spierig Brothers, 2009 / *Stake Land*, Jim Mickle, 2010

Daybreakers is a variation on Matheson's tale, once again featuring humans and two types of vampire: one "medicated" and the other not. Furthermore, it uses vampires as a means of restoring ecological order to the world, but it uniquely features two attempts to do so, with the first attempt unsuccessful, leading to the evolution of its original pest-control agent into something more aggressive.

Daybreakers, not unlike *I Am Legend*, opens with a mysterious plague/contagion whose origin cannot be identified but which quickly reduces the Earth's population to 5 percent of its previous number. The survivors are, by and large, vampires. There are some humans left, but they are now the only food source for the planet's dominant species. This constructs the main human/vampire opposition in the story, but unlike *I Am Legend*, the vampires in this film look exactly like humans, except for their fangs, lack of reflection, their dependence on human blood to survive, and the fact that they never age. Vampires thus emerge as a concentrated, or excessive, form of humanity that constructs them as perfect consumers. Evidence for this contention may be found in the fact that the post-outbreak society has not collapsed but has instead become structured as a hive of offices. This hive structure is partly explained as a means of protection for the vampires, who are extremely sensitive to sunlight, and there is a network of underground walkways and tube trains that serve as a means of transport for the undead clerks—literally Marx's (un)dead labor[74]—to their UV-resistant-glass-walled offices. The work that they perform in these offices is never explained, but all aspects of the previous society seem to have been retained, except food production, and vampires still require money in this post-human world. None of the office staff are shown leaving the city and returning home, giving the impression that most of the vampires' existence is contained within a network that culminates with their office. The only exceptions that exist, or are allowed, outside of that containment zone are the humans, the military that hunt them down, and vampires wealthy enough to own properties beyond the metropolis and a car to reach them—automobiles have blacked-out windows and are fitted with various video cameras so the driver can see the road ahead.

This constructs two very different environments, even ecologies, within the story that delineate the boundaries between the natural and the unnatural worlds—which is made more interesting if the vampires are framed as an ecological force created to restore balance to the world by ridding it of humans. One world is strictly controlled and contained, an interior world that is constantly surveilled by CCTV cameras and based on regulations and never-ending consumption; the artificiality of this environment is further emphasized by the constant focus on synthesizing human blood. The other world is open and unpredictable, food is scarce, and life is perpetually dangerous. It is in this open world that the remaining humans hide from the vampires, exploiting the fact that during the daylight hours the vampires have no protection at all (although later we do see some covered from head to foot in protective clothing), and although the military sends out hunting parties for the humans, the humans also attack the vampires, constructing it as an unstable, unpredictable domain.

Thus, it is not so surprising when a member of the human resistance, Audrey (Claudia Karvan), arrives at the home of Edward Dalton (Ethan

---

74. See Piatti-Farnell 2014, 99.

Hawke)—head scientist for Bromley Marks Corps, who are trying to synthesize a blood substitute—on the periphery of the city to enlist his help. The dangers of living outside the controlled space of production are more graphically shown when Edward is attacked by a Subsider that is hiding in his house. The Subsiders are peripheral figures in the story, but not unlike Matheson's third species, they are ultimately the most important.

The beginning of the narrative suggests that the "outbreak" was directly connected to bats, with quick cuts to flying and screeching images of bats appearing at the beginning and at various points throughout. However, while the bats are responsible for the vampires, they are by no means similar to them. As mentioned earlier, the infection from the bats seems to have done little to change the vampires' appearance and, in many ways, has made them even more human as undead corporate machines—with little need to sleep, never aging, or getting ill, vampires are literally human machines in the workplace. However, when the vampires are unable to procure human blood, their appearance begins to change: their ears begin to elongate, their teeth grow sharper, their bodily hair falls off, and eventually they grow large flaps of skin under their arms that slowly become wings. In fact, they gradually transform into huge bats. In this form they become large, feral, uncontrollable beasts that live in the dark recesses of the vampire world away from the surveillance cameras and even outside of the city. They attack vampires for food and more clearly exemplify a complete breakdown of the "civilized" world that destroyed the planet's ecosystem. In this sense, they would seem to be the end goal for the contagion from the bats, with the vampires serving as the first wave to reduce the population and the Subsiders serving to re-naturalize what is left.[75]

This makes the Subsiders the endgame of the outbreak: to annihilate consumerist civilization and restore the world to a more natural state of ecological balance without human influence, which puts the vampires in a curious position within the film. The human blood they require acts as medication, keeping them from transforming fully, which more closely aligns them to the hybrid human-vampires in *I Am Legend*. In Matheson's novel there is a group of the infected who did not transform immediately and had time to find a concoction that prevents them from turning but does not fully restore their humanity. Indeed, the hybrid nature of the vampires in *Daybreakers* is further seen at the end of the film. Here, Edward has found a vampire, Elvis (Willem Defoe), who has become human again by a process of exposure to sunlight

---

75. This idea of a bat bite turning humans into hybrid, bat-like creature is graphically seen in the earlier film *The Bat People* (Jameson 1974), in which a doctor who specializes in bats gets bitten by one while exploring some caves. He slowly begins to transform and manages to pass the contagion on to his wife. As the film ends the husband and wife become part of a community of bat-people that live in the caves.

coupled with submersion in water before he bursts into flames, which restarts his heart (a beating heart signifies "human" in the film). Edward and Elvis discover that if a vampire drinks their blood, then that vampire becomes human again. In part, this shows that the vampires are not fully turned and can be restored to their former humanity, though as seen above they really have not changed that much in the first place. Furthermore, this process of reversion reveals why the planet requires the Subsiders to become dominant.

At the end of the film, Edward and Elvis have triggered a tidal wave of vampire-to-human transformations—the vampires can smell when humans are near and are so hungry they instantly attack and then become human themselves—which sees the two of them drive off into the unknown future as the sun rises on a new day. However, just as this happens, a large, bat-like form swoops upward across the closing shot, screeching. This bat-like beast is presumably a fully transformed Subsider, a creature that can never become human again and which is the true signifier of the return of planetary ecological balance. This subverts the hopeful ending of the narrative, as the inevitable decline of the vampires will not leave the few remaining humans to rebuild the world but will rather allow for the rise of the Subsiders, who will be more than happy to feed on them.[76] Consequently, the end is not a new dawn for human civilization but the continuing rebirth of the planet beginning its final stage.

Daybreakers creates a parallel version of the present time of writing, which takes place in 2019, but one that is on the verge of consuming itself until the ecosystem steps in. *Stake Land* is based on a similar premise, a near-future where a global pandemic is turning people into vampires. The story centers on a young boy named Martin (Connor Paolo) whose family is killed by vampires, after which he is adopted by an aging vampire hunter called Mister (Nick Damici). Mister and Martin head north across America in search of "New Eden," stopping at small encampments of survivors and avoiding main roads and cities that are largely controlled by The Brotherhood, a fundamentalist militia who see the vampires as the work of God. Unlike in *Daybreakers*, civilization in *Stake Land* has collapsed; cities are abandoned, desolate, dangerous spaces, leaving the survivors to group together in the countryside away from built-up areas. The vampires in this film are feral, violent, unthinking creatures driven by their need for human blood and share few traits with traditional vampires other than their extreme allergy to sunlight. Their wild nature makes them more obviously a product of the environment trying

---

76. This is made even more likely, as leading up to the film's denouement, Edward is shown killing his former assistant who had finally discovered how to safely synthesize human blood, meaning that more and more of the vampires will become Subsiders before they can be turned back into humans.

to reset its relationship with humanity, especially in light of the abandoning of cities and the wholesale disintegration of economic and social frameworks. As Rhonda R. Dass observes, "the hybrid zombie-vampire monsters of *Stake Land* ... [are] sufficient to overcome human resistance, topple civilization as we know it, and leave in tatters the social fabric that once supported it" (Dass 2017, 153).

In many ways, the story follows the blueprint laid down by Matheson's *I Am Legend*, envisioning a world that is broken beyond repair. Indeed, this is a template used by many post-apocalyptic zombie films that describe a world where humans are as monstrous as the undead who prey on them, with little or no sign of resolution or hope, as seen in *28 Days Later* (Boyle 2002), *The Road* (Hillcoat 2009), and *The Book of Eli* (the Hughes brothers 2010). *Stake Land* presents a variation on this theme insofar as the future is constructed as a step back into the past. Consequently, the breakdown of society into small rural groups of survivors, rural militia, and groups of outlaws feels like a journey back to the American Wild West, a time when humanity was more concerned about its survival than the wholesale exploitation of the planet's natural resources for profit. However, this analogy is not without its problems: firstly, as Amanda Hobson points out, surviving the apocalypse in *Stake Land* requires forgetting the past (Hobson 2015, 128), which effectively separates the surviving humans still more from their surroundings; secondly, much of the story of the American frontier is about colonial expansion, particularly in terms of railroads, industrialization and the genocide of indigenous populations. This new vision of the old New World reveals something of an oppositional or devolved view of American history where the vampires return as a traumatic memory of the nation's past, feeding on the "civilization" that previously prospered by consuming its lifeblood, traditions and culture for its own ends. In the same vein, expansionism has descended into isolationism and consolidation, with the land divided up by a series of extremist cults, of which The Brotherhood is the prime example and who facilitated the downfall of civilization by using vampire-filled vehicles and airplanes as biological weapons to decimate cities across America. Indeed, it is almost as if the land has traveled back in time, leaving the humans disoriented and lost in this "new" world. The sudden transition to an environment two or three hundred years in the past means humanity has no other choice but to devolve into an earlier form of societal cohesion, one built on shared values and the will to survive from day to day.

The mise-en-scène of the film accordingly focuses on the landscape: rural countryside, woodlands with wooden houses, overgrown and decrepit roads and railway lines, not unlike the reclaimed city in Lawrence's *I Am Legend*. Unlike narratives such as Matheson's that are set in cities where vestiges of human civilization dominate the story, the humans seem to be minor players lost in the non-technological environment and at the whim of any changes in the world around them. This premise reflects upon the extent of humanity's dependence on the environment and its increasing unawareness and lack of understanding of it as it has been replaced with a "manmade," artificial environment. This point is brought home by the start of the plague, which appears to come from nowhere. The film itself focuses on the world after the initial outbreak and does not mention any cause at all, but a series of webisodes that act as prequels was released to coincide with the film, one of which touches in some measure on the early days of the contagion. The first of these short films, "Origins,"[77] briefly shows a television set playing in the background with a news report showing people being attacked by the sufferers of this new plague/contagion. There is still no explanation of what it is or where it came from, but this film and the others in the series show people emerging from the undergrowth and attacking the living as creatures created by the environment to attack and devour the blood of the species that most threatens it.

Unlike in *Daybreakers*, where the contagion is connected to bats, or Matheson's *I Am Legend*, where the contagion is connected to nuclear radiation and genetic mutation, in *Stake Land* the contagion spontaneously appears. The film suggests, however, that perhaps the contagion has not extended to all parts of the globe, as it focuses entirely on North America. The webisode "Mister"[78] suggests that Mexico has been infected as well, but it describes Canada as the "New Eden." No explanation is given as to why Canada has not been infected—though cold weather is often shown as a deterrent in post-apocalyptic zombie narratives—but as Martin (Connor Paulo) and Mister (Nick Damici) approach the border, they meet a young girl who describes Canada as "the wilderness," almost as if the environment is so untouched or unaffected by humanity that it has no need of vampires to protect or reset its ecosystem. Mister decides to leave the two young people to complete the journey alone, and as they drive through the border, the road is encased in trees, re-envisioning Stoker's tale where the "forest" is the territory between civilization and modernity and the vampire's lair. Here, it is still a place beyond civilization and the need for vampires, no longer a mystical place of transition and transformation but the last sanctuary of safety where humans can once again become part of the ecosystem. Yet, there is no suggestion that

---

77. The film's creators have suggested that they had originally intended the story to be entirely made up of webisodes rather than a movie, though the ones posted in the following link suggest much that links both expressions of the narrative. See "Stake Land—Origins," *Glasseyepix*, 11 May 2011, https://m.youtube.com/watch?v=9EXSQR6JXS4&list=PL34A2B05 A7EDD0627&index=7, accessed 13 June 2019.

78. "Stake Land—Mister," *Glasseyepix*, 11 May 2011, https://m.youtube.com/watch?v=B12P iMFSg68&list=PL34A2B05A7EDD0627&index=6, accessed 13 June 2019.

worldwide balance will be reinstated anytime soon, as the vampires are not going nowhere; they have not been superseded by a more "pure" form of destructive organism, as in *Daybreakers*, nor are they about to be replaced by some new form of hybrid, as in *I Am Legend*. Rather, the undead are here to stay until they starve or are slowly killed by the remaining human community at some indeterminate point in the future. That said, this uneasy peace seems to ensure that humanity will not be exploiting its environment in any large-scale way for some considerable time, thus allowing the Earth to restore itself.

The next examples envision very different futures from those discussed so far, with the ecosystems under consideration exhibiting a more strategic approach in their dealings with the human contagion.

The Time Machine, George Pal, 1960 / *The Colony*, Jeff Renfrow, 2013

The Time Machine is based on H.G. Wells' 1895 scientific romance of the same name, and while it starts in the late Victorian period, it culminates in the distant future. George Pal's film creates its own variation on the book, envisioning the future as a result of humanity's consumption and exhaustion of the Earth and, on some level, an ecological backlash against this. The movie tells of an inventor, H. George Wells (Rod Taylor)—in the scientific romance he is known simply as the Time Traveler—who builds a time machine. He invites four friends to his house in London on December 31, 1899 and mysteriously disappears but not before telling them to come back five days later at 8 p.m. When he returns, his clothes are disheveled and he looks exhausted and proceeds to tell them the fantastic story of his travels 800,000 years into the future.

Of particular note in this story is that he stops the machine in September 1917, June 1940, and August 1966 when Britain is shown to be at war, with each conflict becoming increasingly violent and destructive—the machine does not move in space, so each time it stops it shows the same scene and the differences time has incurred upon it. The first two dates are obviously the two world wars, but the last shows a futuristic society with a monorail running on an elevated track in the background and an "air-warden" wearing a shiny silver suit very similar to cinematic 1960s space travelers—and where an unknown/fictional conflict is taking place. As the film was made in 1960, we can assume this is a reference to the ongoing Cold War between America and the West against the Soviet Union, though no sides are named at this stage. George gets out of his machine to investigate this future world, but a rocket appears in the sky, hurtling toward him. The resulting explosion is nothing the traveler has ever experienced, and it not only destroys buildings but also causes the ground itself to rupture and buckle.

This scene is framed not only as the ultimate example of mankind's violence, which is represented in a series of escalations in each successive pause in the traveler's journey, but also as the end of the Earth's patience with its troublesome inhabitants. The resultant volcanic lava shown bubbling up through the concrete and flooding the streets of London is described as the planet's revenge on humanity. George barely gets to his machine in time as the lava approaches, and he hurtles into the future, visions of the destruction of the world and its gradual regeneration speeding past him together with signs of a new human civilization growing, much smaller than the time he came from. He eventually stops his machine again in the year 802701, over 800,000 years in the future. Looming behind him is a huge, temple-like building with a large door built into it and a sphinx head on top, but everywhere else is lush, green vegetation, a vision of an ecotopian (eco-utopian) future (Hollm 1999, 48).

As George explores further, he happens upon a group of young people laughing and lounging by a river, all seemingly in their early to mid–20s and all blond and without a care in the world. Even when one of their number is swept away by the current of the river, none of them respond. George, however, dives in to save the stricken girl, Weena (Yvette Mimieux), and with her help begins to learn something of this new world. Her people are called the Eloi and do no work, grow no food, and in fact do nothing but live a life of leisure, returning to a large, ruined meeting hall for meals that are mysteriously left out for them during the day.[79] Weena then leads George to another ruined building that appears to be the remains of a museum where she plays him the "rings"—these are metal rings that play a recorded message when spun on a special surface—which tell of the events leading to the current situation on the Earth. The first one George hears says the following:

> The war between the east and west which is now in its three hundred and twenty-sixth year, has at last come to an end. There is nothing left to fight with, and few of us left to fight. The atmosphere has become so polluted with deadly germs, that it can no longer be breathed. There is no place on this planet that is immune. The last surviving factory for the manufacturing of oxygen has been destroyed. Stockpiles are rapidly diminishing. And when they are gone, we must die.

And the next:

> My name is of no consequence. The important thing you should know, is that I am the last who remembers how each of us, man and woman made his own decision. Some chose to take refuge in the great caverns, and find a new way of life far below

---

79. In a curious way, the lives of the Eloi are very similar to the characters in the well-known children's television program *Teletubbies* (Davenport 1997–2001).

the earth's surface. The rest of us decided to take our chances in the sunlight. Small as those chances might be.

There are two points of interest here. The first is the war between East and West, which one can only assume is the one that George had witnessed the beginnings of in 1960. If this is so, then it would mean that the Eloi, and what we discover later are the Morlocks, have been evolving into their current state for more than 800,000 years. As observed by Katalin Csala-Gáti and János Tóth, "what has made the Eloi so helpless and feeble is that they have lived under unchanged circumstances for millennia" (Csala-Gáti and Tóth 2003, 13). This leads to the second point, namely that the two separate groups, while both originally being human, have evolved to look like completely different species. Indeed, the Eloi could be humans and the Morlocks invaders from another world. George had not really seen much of the Morlocks at this stage, so it is worth discussing this encounter before examining the differences and relationship between them and the Eloi.

George discovers that his time machine has been dragged into the building with the Sphinx's head. Unable to gain entry himself, he chances upon a field with low chimneys/vents set into it from which he can hear machinery. He begins to climb down a shaft when an air-raid siren sounds, and Weena, who was with him, begins to blindly walk toward the Sphinx, where she is joined by many other Eloi who, as if in a trance, enter the building, whose doors are now open. George runs after Weena, entering the building himself as the siren stops and the doors close. The Eloi still blindly walk forward, now corralled and guided by the Morlocks that have appeared out of the shadows. The Morlocks are completely different from the Eloi, green-skinned and white-haired, stocky and muscular in stature with large clawed hands and glowing white eyes and seemingly all male (at least the ones seen here). They appear to operate large machines and turbines with no determined purpose—though potentially these could be the oxygen-producing machines mentioned by the rings—but we are shown a cave full of Eloi skeletons, which, at least in George's mind, suggests the Morlocks farm and consume them as a food source. Indeed, there is much about the situation that invokes vampiric texts, such as *Daybreakers* and *The Strain*, where humans are farmed for food. The Morlocks are constructed within a similar framework as creatures of darkness and shadows, extremely sensitive to daylight and possessing large fangs.[80] Matthew Taunton believes this division is a vision of a Marxist utopia where the former class distinctions are reversed and the working, lower classes now literally live off the leisured, upper classes

---

80. The director and writers were probably unaware that traditionally Chinese vampires are green and covered in white hair, which would give them a passing resemblance to Morlocks, though the former hop rather than lope.

(Taunton 2014). However, this can equally be read in terms of ecological sustainability.

Returning to the explanation of the rings, both Eloi and Morlocks were once human, both trying to survive the apocalypse caused by mankind that left the Earth virtually incapable of sustaining human life, even with the artificial attempts at life by way of replication of the natural processes of the planet's ecology. Those that chose to live under the ground still needed to produce oxygen, as they seem to have done with the help of the many mysterious machines that fill the underground caverns. Of course, food would have been in incredibly short supply, suggesting that consuming human flesh became a necessity for the unhealthy or weaker members of the underground society. Over time, as the poisons in the Earth's atmosphere began to subside and the oxygen produced underground began to improve the air above ground, together with the regeneration of the planet's ecosystem a bartering system might have been introduced where the residents of the surface gave "sacrifices" to those underground to continue to provide oxygen and potentially other services as well. This scenario presents not only a self-sustaining humanity but a humanity that fulfills an important role in the ongoing health of the ecosystem as a whole—a kind of ecotopia, as described above. The short chimneys or vents noted by George are not just for the release of heat and smoke from the machines but for the release of oxygen as well. Furthermore, the Morlocks' continued check on the population of Eloi on the surface prevents humanity from having a negative effect on the environment; it is a perfectly balanced system until George arrives.

George of course comes from a time in history when futurity was inextricably linked to capitalism, exploitation, consumerism and the superiority of the human race. Consequently, his influence can only ever work toward the return of the conditions of his age. George attacks the Morlocks in the cavern and even convinces some of the Eloi to turn on their captors, already signaling a change in the established order, before he is forced to escape in his machine from an attack by the enraged underground dwellers. Returning to January 5, 1900, he realizes there is nothing there for him, and after talking to his disbelieving friends, he returns to the future he left to save Weena and the Eloi. However, it is unlikely that many of the Morlocks were killed—one would expect much larger numbers and a network of caverns/communities, given the time they have been evolving. Furthermore, there are no stores or supplies of food for the Eloi to live off, making their long-term survival problematic at best. Thus, the dangerous effects of the past on an ecologically balanced future are likely to be minimal. The next film could be interpreted as filling in the missing part of George's travels to the future, where the Earth has turned against its human population, forcing it underground and dividing it into two separate groups.

Chapter 3. Undead Eco-Warrior 107

The Colony does not specify the distance of its future setting, but another ice age has overwhelmed the world. In the wake of ecological disaster movies like *The Day After Tomorrow* (Emmerich 2004), *Arctic Blast* (Trenchard-Smith 2010), *2012* (Emmerich 2012), and *Snowpiercer* (Bong 2013), *The Colony* envisions a dramatic environmental shift due to the exploitation of Earth by mankind. *The Colony* thereby continues the theme of planetary revenge from George Pal's *The Time Machine*, though this film tries to freeze humankind to death rather than burn civilization into ash via volcanic eruptions and lava flows. In this unspecified future, the entire surface of the Earth is frozen, prompting its inhabitants to migrate into huge underground colonies—the speed of this migration is difficult to ascertain, as one of the characters, Sam (Kevin Zegers), who is in his mid-20s, tells of the collapse of society when he was a young child (early teens?), suggesting it happened in a matter of 10–15 years, yet the underground complexes are huge, and some have enormous and extremely sophisticated structures above them designed to melt the snow and defrost the surrounding ground. When humanity first moved into the "colonies," they were vastly overcrowded, but it soon became apparent that the cold and the lack of food were not the only threats, as a scarcity of medical supplies soon saw influenza becoming an existential threat. The population subsequently dwindled exponentially, leaving most of the colonies with only 50 or so survivors each. This diminished resistance to disease mirrors the end of H.G. Wells' *War of the Worlds*, which also features an ecosystem working against foreign bodies. In Wells' story, Martians invade the Earth and easily overcome all human resistance, but at the point of global domination they succumb to the common cold, against which their alien immune systems have no resistance. However, the scientific romance, and indeed the later film version by Steven Spielberg (2005), suggest something more:

> They were undone, destroyed, after all of man's weapons and devices had failed, by the tiniest creatures that God in his wisdom put upon this earth. By the toll of a billion deaths, man had earned his immunity, his right to survive among this planet's infinite organisms. And that right is ours against all challenges [Spielberg 2005].

This implies a particular connection between an ecosystem and those who are part of it.[81] Although neither the book nor the film explicitly say that this is much more than an accidental, if long-standing, acquaintance, it does suggest that they are inherently part of it, in sharp contrast to the Martians who are not and therefore succumb to its bacteria. This idea of a planetary autoimmune system is quite common within theories around Gaia and a self-regulating Earth, which see humanity's place within it only sustainable if

---

81. The B-movie *Island of the Burning Damned* [Night of the Big Heat] (Fisher 1967) envisions something similar where energy vampires from outer space are killed by a straightforward rain shower.

it becomes part of the solution to ongoing climate destabilization rather than a continual cause of it (Clarke 2017, 209). Consequently, the catastrophic effects of influenza and similar diseases on the surviving human community in *The Colony* can be read as mankind becoming an alien race to the Earth, no longer part of it but an unwelcome intruder.

The film constructs the contagion as an autonomous entity, a physically foreign other that stalks the tunnels and spaces of the colonies, hunting down new victims. Indeed, it is framed as positively vampiric. In the very first scene, a young couple runs down a darkened, claustrophobic passageway, chased by an unseen entity, which then cuts to a scene on the surface where an inhabitant of the colony is questioned at gunpoint to ascertain whether he has contracted the flu—suspected victims are quarantined and, if they come down with the disease, must choose between exile from the community or execution. The effect of this hidden, ever-present disease is to make all spaces unsafe, creating an inherently Gothic mise-en-scène of shadows and imminent danger. Furthermore, this configures space itself as vampiric, where the familiar can suddenly become dangerous and where the ongoing tension virtually sucks the life and rationality out of all those within it. The vampiric form of the virus finds true physical shape in the other "species" that live in the underground complexes and who more closely resemble the Morlocks in their propensity for human flesh.

In *The Colony*, the various colonies try to keep in touch—Colony 7, where the story begins, shares a pledge of protection with its neighbor Colony 5—but Colony 7 has not heard anything from Colony 5 for quite some time, so they send a party to assess the situation. Colony 5 is a two-day hike through the snow, and once the party reaches it, they find a trail of blood leading to its entry door and a hole blasted in the wall near it. They head to one of the large ventilation chimneys—oddly drawing parallels to the vents in the Morlocks' underground complex—and climb down it. Inside Colony 5, the darkness of the interior scene increases—real and metaphorical darkness alike—more directly referencing the horror genre and the sense of entering a vampire's lair: a confusing, nightmarish, vertiginous space where death can strike at any moment.

As they descend further into the complex, the expedition comes across the cavernous space that the vampires/cannibals have turned into a huge charnel house littered with body parts and corpses, some hanging from hooks and chains, while one of their number is methodically hacking his way through bodies on a large, blood-soaked table. Ratcheting up the action, the narrative increasingly references the vampire genre, and as the figure hacking the bodies realizes he is being watched and alerts the others, what seemed to be a floor strewn with dead bodies begins to stir and come alive. It is at this point, and particularly in the figure of their leader, that they change from flesh-eating humans into supernatural vampires—this last is quite important, as they seem to mirror the behavior of the influenza virus in the film: appearing from nowhere, unnatural or superhuman in virility, and seemingly indestructible. The cannibal/vampire leader (Dru Viergever) is a huge figure in a large, almost medieval coat—it is ragged, calf-length and strewn with chains, making it appear as though it is from

Vampires as ecological revenge. Dru Viergever as Leader in *The Colony*. Directed by Jeff Renfrow (Entertainment One, 2009).

some form of barbaric past, connecting it to figures such as Vlad the Impaler (a historical figure often connected to Dracula and vampires in the popular imagination, seen as particularly brutal, bloodthirsty, and associated with a time of plague and other deadly diseases that ravaged Europe).

This vampiric frame of reference is reinforced when the leader opens his mouth to reveal rows of teeth filed down to sharp fangs—all the "vampires" have these—and the primal noises he makes (we never hear him speak) link him directly to the vampires in *30 Days of Night* (Slade 2007), where the undead are shown as soulless, alien creatures as violent and cruel as the arctic weather they seem to have manifested out of.[82]

The vampires swarm after the human intruders and seem relentless in their pursuit. Sam and Briggs (Lawrence Fishburne) are the only members of the party to escape, which they manage to do by trapping the vampires in a lower level of the complex and then destroying the tunnels with dynamite once they have reached the surface. The two survivors immediately set out on a return journey to Colony 7 to warn them of the "vampires," but by the time they reach a large, ruined suspension bridge midway through their homeward journey, the creatures have miraculously escaped and have almost caught up with them. In a struggle with the leading "vampire," Briggs destroys the bridge with explosives, sacrificing himself in the process and

---

82. There is something of a subgenre within vampires that features a more feral variety that can be seen to evolve from Murnau's Count Orlok. These are far more animalistic in appearance and behavior and could never be mistaken for human in the way that Stoker's Dracula was, as shown in films like *Salem's Lot* (Hooper 1979), *Blade II* (del Toro 2002), *Prowl* (Syverson 2010), and *The Strain* (del Toro and Hogan 2014–17), among others.

leaving Sam to make his way back alone. Upon his arrival, Sam tries to warn his colony of the danger posed by the "vampires," but internal struggles lead to his incarceration. He manages to escape and rendezvous with his partner Kai (Charlotte Sullivan) to collect potentially the last remaining seed samples on Earth, heading toward an unknown colony that he learned about while in Colony 5 that has managed to thaw the ground around its complex.[83]

Just as they are about to escape, the "vampires" arrive, somehow manifesting at their door by the will of the environment outside. The undead stream into the facility, not unlike the contagion that decimated the original population, and although the vampires have been shot and seemingly killed, they seem to be as resistant as influenza to any "cure." This point is reinforced during Sam's fight with the leader of the intruders. The members of Colony 7 are gathered in the central hall when the vampires break into the room. Some of them, Sam included, manage to escape through a large vent while another of their number, Mason, sets off an explosion just as the vampire leader grabs him. The explosion fails to kill the vampire but propels him into the venting system just behind Sam. The two begin a struggle and tumble out into a room where the leader repeatedly hits Sam and brings his sharpened teeth/fangs ever nearer to Sam's face. Sam manages to wriggle out of the vampire's grip and finds a heavy metal bar with which to beat the vampire to a pulp. Finally, Sam tires of the struggle and rises from the bloodied body of the vampire to leave, but the vampire has still not died and grabs hold of Sam's foot. Sam grabs a large curved blade, places it in the leader's mouth and pushes until the top of the monster's head is sliced off, reenacting the kind of beheading that kills vampires in much European folklore and literature.

Finally convinced that the leader is dead, Sam collects some seeds as the complex fills with fire and smoke from the explosion. On the surface, the few survivors of Colony 7 decide to join Sam in his search for the unknown colony where human life might finally find hope. As with *The Time Machine*, the narrative suggests that hope may prove false, as not only is the unknown colony a long distance away but there is no guarantee that all the vampires were killed in the explosion and resultant fire, especially given the previous experience of Colony 5 and the escape across the frozen bridge. Furthermore, there is no guarantee that this is the only group of such rabid creatures and that more will not be roaming between the many remaining colonies, killing the inhabitants in tandem with the influenza virus.

The final two films in this chapter show an equally aggressive approach from the planet's ecosystem and one that purposely forces people out of

---

83. It appears that a huge amount of plant life has been destroyed by the permafrost covering the Earth's surface and that the seedlings in Colony 5 might be the only hope for survival left to humanity.

the cities and human-controlled ecologies, not to do away with humanity completely but to integrate it back into the natural environment and, as shown in the final example, into the biological evolution of the flora and fauna itself.

The Silence, John R. Leonetti, 2019 / *The Girl with All the Gifts*, Colm McCarthy, 2016

The Silence is one of three films that came out between 2018 and 2019 that feature mysterious creatures that appear almost from nowhere and force humanity to abandon all population centers, from the biggest cities to the smallest towns, to save themselves. *A Quiet Place* (Krasinski 2018), *Bird Box* (Bier 2018), and *The Silence* all show metropolitan centers as places of danger and often death. *Bird Box* and *The Silence* further depict cities as hyper-sensorial places full of noise and movement that seem to invite the destructive forces of nature into their midst. Both films see their respective lead protagonists fleeing the city as it descends into chaos and hysteria—*A Quiet Place* suggests this has happened, but its narrative centers on the period after the metropolitan overstimulation has occurred and society has already collapsed. Indeed, it is suggested (more obviously in *Bird Box*) that it is this overstimulation, the continual bombardment of images, noises, and sensorial overload, that causes the monsters to appear, manifesting them from the antihuman cacophony that mankind has created in their completely self-centered, consumerist lifestyles of the late–21st century. Further, it is mankind's total separation from the environment that seals their fate, saving those who are still connected to it but condemning those who are not. In this sense, the various plagues that visit human civilization in each film can be seen as manifestations of the planet's autoimmune system specifically hunting out and killing those who are "strangers" to the Earth's ecosystem. *A Quiet Place* sees them as flying alien creatures with hugely sensitive hearing to capture their prey but which reflect their ecological lineage and have heads that curiously seem to feature large petals—not unlike the Demogorgon in *Stranger Things*. In *Bird Box*, the entities are again described as alien in some way—which can equally be explained as coming from a place beyond the human—but are never really seen and seem more like a visual, psychotic dream, overloading the senses of all those who "see" them and causing the victims to go mad. In many ways they could as easily be a concentrated version of everyday life reflected back at the beholder—an equal but opposite excess—causing an affective overload that makes them self-destructive and suicidal. *The Silence* again uses the idea of hearing, but rather than seeing the cause of civilization's downfall as alien, it more directly shows it as a return of the Earth's past—the accidental release

Eco-horror is in the mind of the beholder in *Bird Box*. Directed by Susanne Bier (Netflix, 2019).

of prehistoric creatures from an underground cave—to consume the planet's anti-ecological future.

Of particular interest in all three films is their highlighting of human frailty and the resultant dependence on the environment around them caused by this. In fact, in all three films it is the ability to be connected or attuned to the world around them that saves the respective leads. In *Bird Box* it is blindness and purposeful not-seeing, and in *A Quiet Place* and *The Silence* it is deafness that comes to the fore. While *Bird Box* is actually more about not seeing—the characters wear blindfolds to not be exposed to the aliens/ecological jouissance and eventually find sanctuary in a school for the blind that seems untouched by the monsters,[84] *A Quiet Place* and *The Silence* include deaf characters who are able to block out exposure to the sensorial overload of civilization and wordlessly connect to both their immediate families and the environment around them—in this sense, they reflect Dracula's silent, telepathic connection to his "family."

Much like the start of *World War Z* (Forster 2013), the main characters in *The Silence* seem to be going about their everyday lives in their home in the city—watching television, getting ready for work, etc.—when things suddenly start to spiral out of control. A team of researchers has been investigating an uncharted cave system and broken into a cavern where seemingly thousands of blind, pterosaur-like creatures have been trapped for thousands of years.

---

84. *Bird Box* rather obviously uses ideas from *The Day of the Triffids*: the central idea of not being able to see because of ecological monsters and finding sanctuary in a school for the blind.

Chapter 3. Undead Eco-Warrior 113

About the size of a cat, the creatures are voracious and deadly, consuming human flesh like a shoal of winged piranhas and attracted by the slightest sound. As soon as they escape the cavern, the "vesps," as they are called, attack and kill the researchers and start toward the nearest cities, which serve as beacons of noise and commotion. This enacts a similar scenario seen in many vampire narratives already mentioned where agents (researchers) from the modern world, often looking for unknown resources to exploit, travel to a space representing the metaphorical past, be it Transylvania, the South American rainforest, or underwater caves, that is more closely connected to a time when the planet's ecosystem was in balance and humanity was not around to affect it. Thus, the vesps become both a manifestation of that past and a means for the environment to return to a time before the Anthropocene began. It is not surprising then that the first place they should head toward is the focus of modern, consumerist, civilization: the city.

The vesps as ecological avengers cannot resist the city with its aural, sensorial, and human excess. Indeed, one might argue that it is the very vampiric nature of the city itself "sucking" the life from the environment around it that draws the creatures to it, and it is from this cacophony that the Andrews family escapes to the countryside with their daughter Ally (Kiernan Shipka), who was made deaf by a car accident. Here, the communication skills they have developed to assist Ally, particularly through her father, Hugh (Stanley Tucci), serve to keep them safe over and above the basic task of remaining quiet. This is confirmed in one specific scene as the family leaves the city by car and the vehicle is attacked by vesps. Caught outside the car, Hugh throws a tire iron across the road to attract the vesps' attention and while doing so is close enough to see that the vesps have no eyes and only use sound to track their victims. On returning to the car he instructs everyone that they cannot talk, whisper, or text (mobile devices and the internet still seem to work at this point in the story), basically leaving signing as the only safe way to communicate.

The Andrews family makes it to a farmhouse, but it is not long before they are visited by a group of men who belong to some kind of cult and have all cut their tongues out, seeing the need to be silent as a commandment from God (oddly suggesting that speech is alien or sinful and deafness/signing is almost holy in nature). Using written messages, they demand that Ally be turned over to them, confirming the special status being deaf has for the cult, or they will attract the vesps so everyone is killed. After a prolonged struggle with the cultists, the Andrews realize they are not far enough away from civilization and must find sanctuary away from the lives they used to lead, so they head north to a refuge in the colder climes of North America—as in *Stake Land*, it appears the prehistoric vampires, the vesps, do not tolerate cold weather.

While the ending is rather trite, with Ally pondering if the surviving humans will adapt to living in silence as she has, the vesps have not moved on or disappeared and are probably flourishing in a world that appears to have little defense against them—the film suggests that even three or four weeks after the initial outbreak the best weapons to use against the ecological avengers are bows and arrows. This further implies that humanity will need to adapt to the new "old" order on the planet in order to survive, as the vesps are not dying out or being destroyed anytime soon. Even more so than *Bird Box* and *A Quiet Place* (both of which see humanity remain largely unchanged), *The Silence* infers that mankind will need to adapt to this new world to remain part of it, not least in becoming far more closely attuned to their environment and reducing the many excessive aspects of their former lives. *The Girl with All the Gifts* continues this idea but sees humanity adapting even more to the new post-apocalyptic world, becoming an intimate part of it.

This new world in *The Girl with All the Gifts* is in the near future where Britain has been overrun by flesh-eating monsters called "hungries." These are actually humans who have been infected by a mysterious fungal disease related to *Ophiocordyceps unilateralis*, an insect-pathogenizing fungus mainly found in ecosystems of the South American rainforest that infects ants, changing their behavior to facilitate the reproduction of the fungus that connects to their victim's spinal column and thereby controls their actions. This South American rainforest connection links the story to Matheson's novel (Neville is immune because he was bitten by a bat in Panama) and contemporaneous works such as Justin Cronin's *The Passage* (the infection comes from bats in the South American rainforest) but more specifically the environment itself. Even more so, it replicates the kind of white abyss of unfathomable nature mentioned earlier but replicated in the unexplored jungle, describing it as an unknown, unrecognizable space (environment) where modern civilization is not welcome. This sees the fungus as a direct manifestation of the environment—not least given the importance of the rainforest in helping to maintain the Earth's ecosystem—purposely made to redress the ecological balance and guarantee that the planetary environment remains safe.

The film's narrative shows that the victims most vulnerable to the fungus are adults whose neural cortex cannot fuse properly with the parasitic lifeform (the vampiric fungus), leaving them with no control over themselves so they become mindless zombies until they are destroyed. However, the fungus binds with children and integrates into their brain functions, allowing them certain types and levels of agency, with some able to utilize and direct their new powers for their own ends. The narrative shows a group of such children studied in a government facility—the government thinks that dissecting their brains might provide a cure—that is soon overrun by zombies, forcing the surviving guards and staff to flee to London, where they think they can

Chapter 3. Undead Eco-Warrior

contact a larger military base called Beacon.[85] This group of survivors, led by the leader of the program, Dr. Caldwell (Glenn Close), includes a special hybrid child named Melanie (Sennia Nanua), who seems to have bonded particularly well with the fungus. Helen Justineau (Gemma Arterton), another member of the group who was originally brought on to educate the infected children, has formed a special bond with Melanie, as she sees her as more than an experiment to save humanity. As the group nears London, they encounter more and more hungries, though many of the hungries are standing motionless because the environment is silent and they are only activated by noise.

London is overgrown with huge amounts of unkempt vegetation, recalling Lawrence's *I Am Legend*, though the main danger is the groups of hybrid zombie children that roam the streets. Heading through the center of London, the group comes across a solar-powered military lab, which they move into, as it is airtight and offers a safe place to stay. The BT Tower looms nearby, surrounded by piles of decomposing bodies, which have sprouted vegetation that has entirely covered the building. The vegetation has produced pods that are full of fungus seeds waiting for the right conditions to burst and signal the end of humankind by filling the air with deadly spores—the irony being that the tower was formerly the focal point of communications in London and will now be the means of the fungus communicating itself across the UK.[86]

However, Dr. Caldwell still thinks the only way forward is to dissect Melanie's brain, even though she has saved the group on more than one occasion and proven her own agency and identity beyond the control of the infection. In a telling scene, Melanie asks Caldwell pointedly, "why should we die, so that you may live?" and runs off to the tower, setting fire to the vegetation and causing the pods to fill the air with seeds.

In the closing scene of the film, the air has cleared, and Helen Justineau awakens inside the airtight laboratory. One side is clear glass, and she looks out and sees Melanie beside a large group of the second-generation children sitting on the ground as though in a classroom. Melanie, clearly in charge, looks at Helen and asks her to read them a story, something that the teacher would do when they were in the original compound. Helen declines, but Melanie points out to her that unlike before, they now have all the time in the world for one.

The ending harks back to Matheson, with Helen Justineau framed as the last surviving human in a world that is now controlled by hybrids—humans who have fully bonded with the vampiric fungus. However, Helen is

---

85. The second-generation zombies/hungries were born from infected pregnant women, as the babies chewed their way out of their undead mothers.
86. Many thanks to Hadas Elbar-Aviram for making this connection.

not required to die as Neville was, as her "legend" is not founded on what she has done but what she will do: teaching the children to create their own stories (Bacon 2018a). What is also interesting here is that Helen is kept in an airtight container with no indication of whether this is a temporary or permanent status—even though the air has cleared, the seeds are still on surfaces, there are still hungries roaming around and not all the children are as well controlled as Melanie. Effectively, she is purposely kept out of contact with the new world, one in which humans are now no longer welcome. The original wave of flesh-eaters has caused the destruction of civilization and the ideology that saw it mindlessly consume the environment upon which it depended. The second wave does not reclaim the planet for itself so much as creates the story of a future where "humanity" is a symbiotic part of the ecology.

It is worth noting that the piles of bodies around the BT Tower correlate to the real-life effects of a similar fungus that grows in the rainforest and causes its hosts (ants) to find the right conditions for the organism's propagation and then remain there until they die, upon which fruiting-bodies (seed pods) grow from their head and release the spores. In actuality, the fungus itself is prone to other fungal infections and secondary metabolization, which can protect the host's ecosystem, in this case the ant's habitat in the rainforest. In *The Girl*, the entire Earth is the fungus's habitat, and Melanie and her cohort are intended to protect the Earth, so any further mutations will not be needed—though exactly how the hybrid children might continue beyond their lifespan is never mentioned. The ending then is curious for the way in which the ecosystem deals with the problem of humanity by not just dismantling its economic and ideological systems, as in *Stake Land*, but by modifying humanity itself. This situates humanity, at least in relation to the planetary ecosystem, as a stage in its evolution to something else, with Melanie and her people ushering in a new stage in the evolution of mankind to a more holistic state that reconfigures it as integral to the environment rather than its destroyer. As I have noted elsewhere, the Earth's ecosystem knows that "humanity as it is can never evolve and the future lays in the hands of the monsters, of those who have not become everything that they can already" (Bacon 2018a). In this instance, that becoming is the legacy of Melanie and her cohort.

The Girl with All the Gifts does not entirely do away with humans, with Melanie keeping her teacher Helen as a reminder of their identity as imaginative and emotional entities to enrich or facilitate the ongoing process of regenerating the Earth's ecosystem. However, the last movie in this chapter reduces humanity still further, rendering it an evolutionary byway, or distant memory, on the evolutionary trajectory of total environmental fecundity.

Humanity's days of continually exploiting the Earth might indeed be numbered, but the instinct and energy for life and the power to survive are inherently part of the ecosystem of the planet. The next chapter continues from the idea of exploitation, showing how even these imperatives can be turned toward the survival of the environment.

CHAPTER 4

The End of the End
Consumerism Will Eat Itself

This chapter considers the ways in which industrialization and consumerism become sources of their own destruction while the vampires that they inadvertently produce ultimately assist in repairing/restoring the ecosystem they were trying to exploit.

Scholarship about vampires, post–19th-century ones in particular, reads/interprets them as an obvious manifestation of consumerism, namely the insatiable consumer that must possess and consume until there is nothing left.[87] Indeed, some of the examples discussed below seem to conform to that reading, showing the ultimate effects of such an all-consuming approach on the world and giving form to the notion of never-ending consumption as a form of immortality. Equally, some of the films here also show how these same vampiric manifestations can actively work against such an ideology—the same one that brought them forth—breaking down the forces of consumerism and actively working to restore ecological balance. The place of humanity within these visions necessarily reveals the way in which a world governed by the rules of consumerism will, quite literally, eat itself.

Ganja & Hess, Bill Gunn, 1973 / *The Omega Man*, Boris Sagal, 1971

Ganja & Hess and *The Omega Man* both carry very strong overtones of racial tensions in America, particularly in regard to the African American

---

87. Voltaire notes in his philosophical dictionary: "We never heard a word of vampires in London, nor even at Paris. I confess that in both these cities there were stock-jobbers, brokers, and men of business, who sucked the blood of the people in broad daylight; but they were not dead, though corrupted" (Voltaire 1927, 144). Franco Moretti says of Stoker's *Dracula*, "Capital is dead labour which, vampire-like, lives only by sucking living labour, and lives the more, the more labour it sucks" (Moretti 2005, 192).

community, although *The Omega Man*'s very specific view of Whiteness rather complicates this. However, they can also be read in terms of a return to an earlier time, one where humankind was more dependent on, and closer to, its environment. *Ganja & Hess* conceals these details in plain sight while more obviously addressing the place of African Americans in contemporary White American society.

Gunn's film centers on Dr. Hess Green (Duane Jones), an anthropologist and geologist who is curating an exhibition at the local museum on ancient African Myrthian culture.[88] Green lives a luxurious life of wealth and leisure in a large mansion with extensive grounds. The house itself is expensively furnished with antiques and works of art, and there is even a butler. In many ways he exemplifies the pinnacle of African American integration into White capitalist society, living a life of luxury and leisure that allows him to indulge in his own personal intellectual pursuits. In part, this examples how Black culture has been consumed by White America and subsequently forgets its own but also shows the beguiling, glamouring nature of consumer culture itself that consumes everyone and everything. It highlights the objectifying nature of consumerism, as explicitly shown by Green himself. He lives a life full of constant good taste in clothes, food, fine wines, etc. and even in his work and home surroundings, from interior decorations to objets d'art. His work at the museum also shows this by treating ancient relics as objects to appreciate and be put on pedestals and admired and whose past and cultural/historical settings are just a means to increase the popularity of the exhibition, changing the aura of ritual artifacts for decorative reproduction.

This feeling is extended to all parts of his life, seeing him detached from his own existence and looking upon it as an object to be appreciated, which similarly extends to his own cultural background, allowing him to pick and choose the one he prefers as opposed to one that he is intimately and emotionally connected to; thus, he takes home items from the exhibition as decorative pieces to make his home look more attractive. One object, though, a Myrthian sacrificial dagger, is not so compliant, and its vampiric nature changes Dr. Green's relationship to everything.

Green's African American assistant, George Meda (Bill Gunn), comes to the mansion to assist in the preparations for the upcoming exhibition. That night, Meda gets increasingly agitated and tries to take his own life, but Green manages to save him. However, a short time later, Meda attacks his employer with the sacrificial dagger, cutting him and then killing himself. Green survives, but when he finds Meda's body, he is compelled by an overwhelming

---

88. This is a made-up name for the film but indicates an ancient Nigerian race and culture that pre-dated the start of the colonial slave trade that began in the 15th century.

impulse to drink the dead man's blood, signaling a crucial change in himself and his connection to the environment.

Curiously, Meda had already experienced the troubling effects of objects, and he had begun to feel the connection to the past land of ancient Myrthia, which was steadily making him more and more agitated. Indeed, it was the dagger itself that compelled him to strike Green with it so that another could be connected to this world of the past. Once Green has been "bitten" by the blade—for that is effectively what it did—the world is suddenly one of the senses rather than of the intellect, and he is now intimately joined to it. This is largely shown through his need for blood, which is no longer the aesthetic, sophisticated experience epitomized in his dinner parties but a guttural, sensual, erotic experience. This connection to the world around him is further reinforced by the arrival of Meda's wife, Ganja (Marlene Clark). She comes to the mansion looking for her husband, but it is not long before Ganja and the newly aroused Green become lovers. Green's life, formerly one of refinement and judgment, becomes one of continual sensual experience and excess but one that slowly takes him back to the past—not his own history, it should be noted, but more of a mythic, African one linked to the fictional Myrthia tribe featured in the museum exhibition. This is shown when Green asks Ganja to marry him and they perform a ritual in the grounds surrounding the mansion, where they only wear robes of wrap-around tribal/ethnic fabric—simple black and white woven fabrics—and share flower blossoms as they proceed through the gardens. Much of this reveals a fundamental linkage to the land and the environment, with their union being part of that, as is their need for blood, which is constructed as a direct opposite to gourmet cuisine as representative of modernity and consumerism.

The shift in time and ecological relationship is further expressed through Green's increasingly vivid visions. These feature the sudden eruption of images from this mythic past that burst into the present, where African figures in ceremonial dress overlay the mansion and its grounds. These allude to a shared African past that has been forgotten, or repressed, because of White American consumerism. This savage but ecologically balanced past is revealed through the dreamlike visions of ancient ceremonies that celebrated the Myrthias' connection to the land, with which they live in union by the drinking of blood, a bond which carries on forever and bestows immortality to those who become part of it. The ceremonies, incorporating the sacrificial dagger that "bit" Green, see him as both sacrifice and consummation of the ritual. He too is now re-connected to the environment and a way of life that is integral to the ecosystem, one which promises to continue forever.

Ganja has now been initiated into this community, intimating how this vampiric/ecological past can begin to establish itself in this consumerist present, and she seems more at home within it than Green. In fact, his indoctrination into White American ideology (the religion of consumerism) seems to be far more deeply embedded than Ganja's, to the point where he begins to be pulled slowly back into it. This "back-sliding" becomes centered around religion in the film, partly reflecting Matheson's theory of cultural symbolism in which dominant social narratives create real physical effects in the "believer's" body. As such, Christianity is shown to embody the lifestyle that Green formerly led,[89] where earthly pleasures and a deep, reciprocal connection to one's environment are disavowed for the spiritual and transcendental. The Myrthian culture, in contrast, is centered on the here and now and a symbiotic relationship with the land that is consummated by blood as a symbol of life (as quoted by Renfield in Stoker's *Dracula*). Consequently, when Green feels he can no longer deny the consumerist life that formerly consumed him, he tries to symbolically return to it by standing before a large cross.

As with Robert Neville, there seems little reason why this should have any effect on the Myrthian vampire inside him, as it should have no connection to Christianity. However, it can be read in terms of the cross as symbolic of the White civilization that took so many Africans (as symbolized by Myrthia) as slaves, effectively "killing" the "old" religions—though this would be a rather simplistic reading, given the various forms of hybridization that occurred. The scene can further be read as expressing the incompatibility of consumerism and ecological balance, where Green has been fundamentally changed by his time spent in the Myrthian past—not unlike Lena in *Annihilation*—so that when he tries to return to his old life, he is unable to because it will no longer accept him. His "DNA," the essence of who he now is, has been irrevocably altered so that the Christian cross is anathema to who he now is; equally, he is no longer "safe" in the eyes of the White culture that he had previously been assimilated into and is now "dead" to that, being a dangerous, black outsider. This incompatibility leaves him trapped between the past and the present, as he is no longer welcome in either world, leaving him no option, "psychologically," other than to stop existing. In contrast, Ganja ends the film walking along the sand dunes not far from the mansion. She is content with being at one with nature in a world where civilization, technology and consumerism mean nothing and happily walks into the future.

The Omega Man also centers around religion and race in its vision of an ecological past in the future. *The Omega Man* is a very idiosyncratic adaptation of Matheson's *I Am Legend,* portraying the vampires more as religious zealots than undead monsters. The story is set in 1977 after a border conflict between China and the former Soviet Union ignites biological warfare that produces a worldwide plague. Military doctor Robert Neville (Charlton

---

89. See Weber 2010.

Heston) vaccinates himself with an experimental serum, making him the only person immune to the contagion. The plague victims, however, do not seem to die but become albino and extremely sensitive to daylight—not unlike Matheson's vampires—with very little blood-drinking. In Los Angeles, where Neville is based, the victims have become part of a cult called "The Family"[90] that aims to destroy all forms of technology, which it holds responsible for the warfare and resultant plague.[91] The leader of the cult, Matthias (Anthony Zerbe), is shown as an oppositional figure to Neville, even though in many ways they configure a very similar view of Whiteness—both are constructed as "White saviors"[92] for their respective worlds, not least in their ongoing battle to gain control of the African American body as seen specifically in the figures of Lisa and Ritchie (to be discussed later)—in their respective approaches to technology and capitalist ideology.

Neville, as a scientist and a military man, embodies exactly the kinds of forces that brought about the apocalypse in the film, which arose from Cold War tensions that threatened mutually assured destruction (MAD) as an ongoing means of political engagement. This framework of aggressive posturing has little regard for the environment beyond national borders and sovereignty, with nuclear proliferation and bioweaponry the more obvious indicators of the perniciousness of such an ideology to the ecology. Neville still manages to embody this idea once civilization has collapsed, as indicated both in his "ownership" of the city and the protected space in which he lives. The city itself appears untouched, with the apocalypse signaled only by an excess of waste blowing around the streets. In all other respects, it just seems abandoned. Indeed, it is as though the ideology of MAD is a framework left perfectly in place but with no humans to enact it. The city perfectly symbolizes that impression, seemingly untouched and just waiting for human actors to once again inhabit it and perform its neoliberal choreography in the designated zones. Neville drives around the abandoned city in a bright red sports car with the top down, toting a machine gun just in case he encounters any of The Family while he is out looking for supplies—Neville is only engaged in his own self-preservation in this narrative rather than actively seeking out and destroying vampires.

Neville's aggressive ownership of the metropolitan area, taking anything he wants while carrying an automatic weapon, perfectly encapsulates the world that created him and which he continues to embody—a world devoid

---

90. The choice of name for the community of vampires is an interesting one because the infamous Manson Family—a commune of about 100 led by Charles Manson—had been based in Los Angeles in the late 1960s, culminating with the murder of Sharon Tate and four others in 1969. See also Ransom 2018, 101, 109.
91. *Stake Land* reflects something of this in The Brotherhood.
92. See Hughey 2014.

of vegetation or greenspaces that is only there to fulfill his needs.[93] His home manifests his aggressive masculinity still more, as it is literally haunted by the ideology of the pre-apocalypse world. Neville lives in a townhouse with an inbuilt underground garage that is heavily defended against the nightly visitations of The Family, enacting in microcosm the problems that caused the downfall of humanity. Within the house he keeps a generator running to power his living space, which contains a television, a record player and all the equipment in his working laboratory where he is trying to recreate the vaccine that cured him. In fact, he does everything he possibly can to maintain a life as closely resembling the one he lived before, with his books, antiques, pictures, and chessboard next to a bust of Julius Caesar as an ever-present specter of the past, a vampiric presence that is waiting to be resurrected to continue sucking the life out of the world and a ghost that Neville is doing everything he can to bring back to life in his laboratory.

In contrast, Matthias is doing everything he can to dispel this ghost and get rid of the past forever. Not unlike Neville, he had a clearly delineated role in the old world, starring on television news and commenting on the developing apocalypse. His response was very different, however, and he firmly lays the blame for the destruction of the world on technology—the ideology behind it rather than the people or the specific machines/weapons themselves—and vows to rebuild the world without it; Paul Meehan notes The Family are more "interested in burning books and trashing museums than they are in drinking Robert Neville's blood" (Meehan 2014, 77).[94] In this way, the relationship between the two men more closely resembles that of Count Dracula and Van Helsing from Stoker's *Dracula*, where it is not so much about good versus evil, although it tries to suggest that, but is more about who is the master—a point mentioned by both characters in Stoker's novel—and for Neville and Matthias, it is centered on who is the master of the city.

Neville's experience of the city is all about himself and the kinds of individuality and gratification that consumerism promotes so that his constant patrolling of its streets in his sports car becomes a way to remember that previous life as well as a way to bring it back: by finding equipment to continue his search for a serum. Matthias observes this about his rival, saying that Neville has "…nothing to live for but his memories, nothing to live with but his gadgets, his cars, his guns, gimmicks…" (Sagal 1971). He further comments on the vacuity of Neville's life, both previously and now: "You see, none

---

93. Ian Cooper interestingly notes that Neville reflects Heston's own conservative views, and the vampire plague represents the failure of the 1960s hippie dream with The Family as its mutant apotheosis. See Cooper 2018, 119.

94. Not unlike the kind of anti-technology ideology put forward by the Luddites, English workers in the early 19th century whose livelihoods were threatened by the introduction of new machinery in the wool and cotton mills.

of it was real. It was illusion. Your art, your science, it was all a nightmare. And now it's done. Finished." The Family purposely eschew all the trappings and luxury that Neville hordes so assiduously and live more of a communal lifestyle, much in the way of a religious community or worker's cooperative, though it is never made clear whether they have any requirements, such as food or drink, other than a place to protect themselves from the sunlight during the day.

In fact, there is much about The Family that echoes the Morlocks from *The Time Machine* in terms of a kind of class division between themselves and Neville; they live in the darkness, are all dressed similarly in black robes, have pasty white complexions—often with lesions—and have white, almost glowing eyes, while Neville lives in his townhouse surrounded by art, sculptures and antiques, drinking scotch and taking pot-shots at the "vampires" while he wears a velvet dinner-jacket and a shirt with a ruffled collar and sleeves. This makes purposeful reference to the readings of Stoker's novel highlighting the decadence of an outdated aristocracy, complicating the designation of who is the real vampire in the story. Franco Moretti sees the vampire embodying a decadent aristocracy feeding off of the youth of the nation and capitalism exploiting the undead labor of the working classes, which tends to construct Neville as the undead monster. However, The Family's similarities to Wells' Morlocks see them signify the faceless, mindless underclasses and workers with closer links to manual labor and their environment, connecting them to the kinds of authenticity that so appealed to philosophers such as Walter Benjamin and Martin Heidegger.[95] Alongside this framing of The Family is that even African Americans become white with albino eyes—*The Omega Man* was released just before *Blacula* (Crain 1972) and *Scream Blacula Scream* (Kelljan 1973), which are both set in Los Angeles and make obvious reference to the ongoing racial tensions in the city occurring at that time. *The Omega Man* in comparison has a more ambiguous approach to the racial tensions in Los Angeles and America in general.[96] There are several black characters in the film, but the ambiguity toward the African American community is most clearly seen in the figure of Ritchie (Eric Laneuville), whom Neville meets after being saved from the "vampires" by Dutch (Paul Koslo), the leader of a group of human survivors.

Not all humans have completely succumbed to the plague yet, and Dutch and Lisa (Rosalind Cash), an African American woman, are in charge of a group of young adults/children who are only just beginning to show symptoms,

---

95. It should be noted that we are never shown The Family undertaking any kind of manual labor within the film.

96. Adilifu Nama reads the film as being demonstrably white supremacist in its construction (Nama 2008, 52), while Xavier Mendik sees it as more ambivalent, if not contradictory, in its approach to race (Mendik 2002, 38).

such as becoming increasingly sensitive to sunlight. One of the children, Ritchie, is clearly beginning to turn into a vampire, so Neville tells them of the cure he is developing, and he and Lisa take the young man back to his house in the city. Once there, Neville uses his own blood to make a serum to cure Ritchie, and we visibly see the boy change from pasty white back to his original color as the vaccine begins to work.[97] Ritchie is keen to try the serum on all those affected by the plague, even The Family, but Neville refuses, saying that they are too far gone to be saved. Distraught, Ritchie runs away to try and help the vampires himself and dies as a result. This marks a curious subplot in the film where Neville and Matthias, the two White "masters" in the film, battle over the African American body. Once infected, Ritchie begins to look like the other members of The Family, who envision something of a worker's commune where everyone looks and dresses the same, but they all have completely white skin, hair and eyes. By accepting the serum from Neville—a form of holy communion, given Neville's Christ-like status at the film's end—Ritchie would retain his own identity but would always be seen as a non–White other. As such, Ritchie's death is almost inevitable, as The Family has no place for saviors or leaders other than Matthias, and to accept the serum would mean accepting the sanctity of Neville's blood sacrifice.

Part of this ambiguity also speaks to the attempts by Neville to restore neoliberal, consumerist principles where the wealthy can economically afford to be part of the consumerist ideology but the poor cannot, with Neville and the surviving humans representing those that have and The Family representing those that have not. Indeed, Neville's place in this hierarchy, and also that of Dutch as leader of the group of survivors, is that of the very wealthiest white men at the very top of the tree who control all of those beneath them. In this sense, the individual identity offered to all those who are saved by Neville's serum (literally his blood) is the illusory kind offered by possessions, technology, and "gadgets." In contrast, the plague strips away material illusion and reveals the fundamental sameness of humanity: a world where everyone is equal and reliant on the environment. The ending of the film affirms this, with Neville killed by a spear and dying in a very Christ-like pose—not unlike the one used in the earlier *Last Man on Earth*—but not before passing on his blood to the faithful so that mankind might survive in his image. This very Christian symbolism harks back to *Ganja & Hess*, where it represents the religion of the slave-owners and the rich, white capitalists who exploit others for their own gain, shading the end of the movie as a potential triumph for the memories/ghosts of the past that might once again find form to ruin the planet that sustained them.

---

97. The film also boasts one of the first interracial sex scenes in an American film, between Charlton Heston and Rosalind Cash.

The next two films look more closely at the objects that focus the all-consuming ideology of consumerism, specifically one item that seems to encapsulate much of what is wrong with mankind's relationship to the wider ecosystem: the automobile. As mentioned above, Neville uses many cars to establish his possession of the city and to maintain his ongoing project of acquisition and consumption to revive consumerist ideology in the post-apocalyptic world.

Christine, John Carpenter, 1983 / *Blood Car*, Alex Orr, 2007

Christine more explicitly shows the vampiric nature of humanity's dependence on fossil fuels and the machines that consume them, explicitly linking "mankind's" sense of identity to his ability to consume natural resources. The film expresses this idea of excessive consumption from its very beginning, presenting a car production line in Detroit in 1958 where the Plymouth Fury was made. The cars are large and opulent, from an era when efficiency and economy were not of importance in car design, and the Fury consumed fuel thirstily.[98] The cars on the assembly line are all in beige and chrome, except for one that is blood red. The line stops for some final hands-on checks, and a mechanic comes over and lifts the hood of the red car, resting his hand on the edge while looking under the front of the car. The hood suddenly slams down, slicing off the man's fingers. Later that same day, another worker enters the car to check over the interiors. He is smoking a large cigar, and some of the ash falls onto the plastic covering the front passenger seat. Obviously, the car, named Christine, is not impressed by this, and as the man turns on the car radio, the camera pans away, only to return at the end of the day when the man is dead, still upright in his seat. As indicated by these opening scenes, the car itself is vampiric[99]—a spirit or demon does not possess it, as seen in *I Bought a Vampire Motorcycle* (Campbell 1990), nor does an extraterrestrial entity, as in *Hybrid* (Valette 2010)—revealing that the ideology itself that drives production is inherently deadly.

This figure of car-as-vampire is developed further once the car finds its first owner—the car consistently selects the weak, the easily influenced, and the impressionable, and just like Count Dracula, it glamours its victims to its

---

98. The car in American culture is of particular importance both as a symbol of modernity and American manufacturing prowess and also as an expression of the American Dream to explore and conquer the huge spaces of the nation, as seen in texts as varied as Jack Kerouac's *On the Road* (1957), *Back to the Future* (Zemeckis 1985), and *Mad Max: Fury Road* (Miller 2015).

99. *The Car* (Silverstein 1977) suggests a similarly vampiric car.

Chapter 4. The End of the End

Deadly desire in *Christine*. Directed by John Carpenter (Columbia Pictures, 1983).

will, slowly turning its driver into an extension of itself.[100] While we never see the first owner, we hear the opinions of others—first the owner of the garage where Arnie, the current owner in 1978, stores Christine:

> Will Darnell: I knew a guy had a car like that once. Fuckin' bastard killed himself in it. Son of a bitch was so mean, you could've poured boiling water down his throat and he would've pissed ice cubes! [Carpenter 1983].

This point is further elaborated by the first owner's surviving brother, George LeBay, while talking to Arnie's friend Dennis:

> GEORGE: Either you're dumber than you look, or you don't know your friend very well. He had the same look in his eye that my brother always had. Probably the only thing my brother ever loved in his whole rotten life was that car. No shitter ever came between him and Christine, if they did ... watch out! He had a five-year-old daughter choke to death in her ... he wouldn't get rid of her. He just rode around with the radio blaring, not a care in the world except for Christine. Only time I ever interfered with it was when Rita killed herself.
> DENNIS: Who's Rita?
> GEORGE: His wife! He didn't care a rat's ass about her! She died the same way he

---

100. By reading the car as Count Dracula, its owners/drivers then represent Renfield-type characters, and more so the Håkån figure from *Let the Right One In*.

did ... then I made him get rid of it ... for decency, ya know? Of course, the car came back three weeks later.

DENNIS GUILDER: What do you mean "came back"? [Carpenter 1983].

It is not always clear whether Christine kills people for energy—it certainly seems to run on fuel rather than blood, for instance—or for more emotional reasons (jealousy, revenge, or the need to be loved), but it certainly stores an amount of life-force, most obviously from human victims, to repair itself or influence objects or people near it. At any rate, Christine is a grotesque representation of the effects of consumerism on those who come under its influence and begin to "love" objects and things—Neville's "gadgets"—over the people and the world around them. In fact, as the story continues Arnie forgoes or loses his need for sex as it is satiated by his love for Christine, further highlighting the autoerotic delights of consumerism.

These effects can be very dramatic, and Arnie himself begins an arc of change similar to Christine's first owner. When we first see him, he is the nerdy, poor kid at school with his glasses held together by tape and constantly picked on by bullies, but as Christine starts to glamour him, his clothes get smarter and he wears a leather jacket, he greases his hair in a backward wave and he no longer wears glasses. His skin, however, concomitantly appears paler than usual, with deepening pools of darkness around his eyes, almost as if the car is feeding on Arnie even as it binds him to itself.[101] Arnie has a girlfriend, Leigh, the new girl from his class at school, but as the jealousy of consumerism begins to take effect, he experiences a surge of desire for individual gratification above and beyond any other attachments, which includes all aspects of the world beyond that relationship. Bruce F. Kawin describes the car's attachment to Arnie as a "jealous crush" (Kawin 2012, 88), but it is more consuming than Kawin suggests, as both Arnie and Christine maintain their identities and sense of self (value) through the "love" of the other.[102] It is no mistake that the car is strongly identified as feminine and that, as their relationship intensifies, Arnie becomes increasingly masculine in appearance and behavior.

Consequently, Christine tries to get rid of Leigh when Arnie takes her to a drive-in movie. One of Christine's windshield wipers sticks, so Arnie gets

---

101. Something of this is seen in *The Red Violin* (Girard 1998), in which an antique violin both inspires and drains the life out of all who play it. Count Orlok in *Nosferatu* actually loves his victim Ellen and wants to devour her because of that; here, Christine is the infatuated vampire. See Christie 2015.

102. Christine's pursuit of Arnie brings to mind a quote from Sheridan Le Fanu's *Carmilla*: "the vampire is prone to be fascinated with an engrossing vehemence, resembling the passion of love, by particular persons. In pursuit of these it will exercise inexhaustible patience and stratagem, for access to a particular object may be obstructed in a hundred ways. It will never desist until it has satiated its passion and drained the very life of its coveted victim" (Le Fanu 2003, 246).

out of the car to fix it. Leigh begins to choke on a mouthful of hamburger and suddenly all the doors lock shut, preventing anyone from leaving the car or entering it to help her. Leigh forces the door, and a bystander rushes in and saves her by performing the Heimlich maneuver on her. Leigh then refuses to go anywhere near the car again, effectively saving herself, but the school bullies are not so lucky. They break into the garage where Arnie keeps Christine and vandalize it, leaving it virtually destroyed. Arnie is distraught, but as he turns for a look at its flank/trunk, he notices a part of the engine is suddenly repaired and looks brand new. Arnie backs away and motions for Christine to continue. The car flicks its headlights on and proceeds to repair itself in front of him—there is something oddly intimate in this scene, suggesting sexual gratification/intimacy within the realm of consumerism and desirable objects, which can be eternally new at the expense of the surrounding environment. Once restored to its original condition, Christine begins to hunt the various members of the gang and kill them, variously running them down, destroying a gas station, or crushing them against a wall, often causing extreme damage to it(her)self. Every morning, however, Christine looks brand new again. Arnie is now in thrall to the car, and even at the story's denouement as he lies dying, impaled by a large piece of glass, he looks happy, caressing Christine's bumper. Dennis and Leigh, who have implemented a plan to destroy the car, successfully maneuver it into a metal crusher, and as the film ends, Christine is pulverized into a cube of metal, but even as the image fades, a piece of metal slowly begins to straighten. While the environment is never explicitly mentioned in the narrative, the effects of such a car from an age that guzzles such a large amount of fuel can only symbolize a world hell-bent on exploiting fossil fuels to the detriment to the planet's ecosystem, as observed by Tony Magistrale: "The excretion or expulsion imagery in the film represents the linear nature of the energy investment; indeed, there is no restorative exchange at all, but only ingestion resulting in polluting effluvia that cannot be escaped" (Magistrale 2008, 58). This further suggests an inescapable and deadly "love affair" with consumerism that will never end, forever consuming and polluting our only home. However, the next film makes these connections explicit, likewise using a car as a symbol for consumerism.

Blood Car is a low-budget film that more explicitly deals with consumerism and its ideological implications and institutional backing. The story is set in a near future in which petrol is hugely expensive and focuses on an inventor, Archie Andrews (Mike Brune), who is trying to make an engine that runs on wheatgrass—the similarity between the names Archie and Arnie is probably not coincidental. However, things do not quite go to plan, and he accidentally constructs an engine that will only work with human blood rather than wheatgrass. Archie's daily routine includes stopping at a vegan stall to buy jars of wheatgrass with which to experiment in his apartment. A stall selling slabs of meat stands directly across from its vegan counterpart, run by a scantily clad girl, Denise (Katie Orr), who offers sexual favors for road-trips in any kind of motor vehicle. Archie's experiments are not going very well, and his test engine refuses to start, no matter how much wheatgrass he puts through it, until one day he accidentally cuts his hand and blood gets into the machine, causing it to instantly spring to life.

Archie pours more and more of his own blood into the engine, which is now in a car and which attracts the attention of Denise. Hooked into the increasing cycle of consumption and desire, Archie begins to trap and kill animals to feed the car—he has adapted the car so that corpses loaded into the trunk will be "consumed" and fed into the engine—but these fail to fuel the engine, and he realizes only human blood will satisfy it. He happens to discover his elderly neighbor dead on the outside porch, so Archie wraps the body in sheets, drags it to the car and throws it in the trunk. The car literally roars into life as he drives to pick up Denise and they have sex. This rather obviously connects some fundamental aspects of consumerism: desire, sexual gratification, objectification, consumption, and of course death. This connection is strengthened as the film reveals a never-ending cycle that takes Archie further and further away from himself—no longer the ethically and ecologically minded vegan he once was but now a cold-blooded killer who will do anything to ensure his own gratification. A seminal moment in this transition is when the girl who runs the vegan stall, Lorraine (Anna Chlumsky), finally persuades Archie to take her out on a date. They go to his flat/house, but Archie has to leave her there alone as he sorts out an issue with the car. Left on her own, Lorraine slips and falls in the shower, breaking her back and leaving her paralyzed, but Archie never returns for her and chooses Denise as his girlfriend. Their relationship is based solely on Archie's possession of a car—a very rare thing in his near-future world except for the very rich—and Denise performing increasingly bizarre sex acts on him. Lorraine represents not only the "good girl" of the story but also the responsible and ethical way to relate to one's environment, while Denise represents the increasingly fetishized and selfish quest for individual gratification.

Unbeknownst to Archie, he has been under observation by government agents since the beginning of the film. The agents are content to simply observe his actions until the car reaches prime efficiency on its special fuel, at which point they decide to step in and requisition the car for the government. During his first altercation with them, Archie manages to fight back, push one of the agents into the trunk of the car—where he's processed—and escape. Later in the film, however, the agents offer him a deal he cannot refuse. If he can make more copies of the blood car, he will have earned any position in society that he desires, all the way up to president. Showing the extent to which Archie has been glamoured by vampiric consumerism, it never even crosses his mind that he could use his potential new position for good and save the ecology he was previously so keen to protect; he only sees this offer as a path to adulation and personal wealth. The film draws to a close as Archie stands, dreaming of his possible future, a dream that is intercut with images of all his former acquaintances—Denise, Lorraine and all the children he teaches at school—disposed of by government hitmen so that no one finds out about his past.

This last part of the film reveals the more structural side of society's destruction of the planet to satisfy its own individual desires. This destruction is hastened not only by global corporations but by governments as well, revealing that the ideology underpinning this destruction is endemic to contemporary society. Not least, and particularly relevant for America's current Trump administration, it explicitly points to the connection of powerful corporations and the government and the inherent corruption behind the political system, particularly regarding its decisions regarding climate change and the environment. The film further intimates the depths of corruption to which consumerist ideology has plummeted, particularly when driven by neoliberal ideals that largely quantify worth in financial benefits and returns. Archie explicitly draws attention to the corrupting influence of consumerism at each stage of the story, as his attempts to find an eco-friendly fuel are skewed from the very beginning. Although he begins by testing the wheatgrass fuel in a simple motor on a table in his apartment, as soon as it begins to run on blood, he moves the testing to a black sports car, a vehicle that has eco-credentials as dismal as Christine's. As Archie gains Denise's favor through his car, he willingly kills to stay with her, which then renders Denise herself his object of desire, which Archie needs to continually pay for to receive the same levels of gratification—the withdrawal of gratification for non-payment is dramatically shown when the car runs out of fuel and Denise immediately stops giving Archie a blowjob and gets out of the car. The film thus follows a pattern of ongoing exchange, correlating and fetishizing financial and sexual rewards. By the time the government agents intervene, Archie has lost all moral bearings and, not unlike Arnie before him, has become a different person and as much of a vampire as the car itself, if not more so. Indeed, as the film ends, the car is no longer seen, and it is Archie who is shown as the true source of greed and corruption, with his choices leading directly to the deaths of everyone he knew, suggesting the extent to which our own individual decisions affect the environment around us.

Both *Christine* and *Blood Car* use automobiles to represent the consumerism of society, but the next films make that aspect even more explicit in their respective narratives, showing a society that is guided and controlled by consumerist desire.

They Have Changed Their Face, Corrado Farina, 1971 / *Snowpiercer*, Jon-ho Bong, 2013

They Have Changed Their Face is far more focused in scale but is spread much further throughout history, hinting at the undying nature of human greed and humanity's disregard for its environment, told through a narrative loosely based on Stoker's *Dracula*. The vampires in this film are a secret society, not unlike the idea of the Illuminati,[103] that has continually driven civilization toward an all-consuming capitalism. The vampires are represented as immortal beings that have moved from body to body over centuries but with little regard for the planet other than as another resource to exploit.

The story is set in Turin, the industrial capital of Italy and home to the nation's car industry—not unlike *Christine*'s opening scenes set in Detroit, which served as a similar metaphor for American consumerism. The main character, Alberto Valle (Giuliano Esperati), is a mid-level employee at a successful car manufacturer called Auto Avio Motors. Alberto is unexpectedly invited to visit the mountainside villa of the company's owner, Giovanni Nosferatu (Adolfo Celi). Alberto's life then begins to change drastically. Not unlike Arnie after meeting Christine, Alberto is taken further and further away from his former life, becoming ever more vampiric as the narrative unfolds. However, one may equally argue the converse, whereby Alberto comes into his own as he is no longer just a cog in the capitalist machine but one of the influencers of its ongoing evolution, or to paraphrase Roberto Curti, he becomes part of the vampiric "haute bourgeoisie" and no longer one of the dominated "lower classes" (Curti 2017, 4).

If the world Alberto left exemplified streamlined industrial efficiency typified by utilitarian modernist design, Nosferatu's villa is even more so, with typical 1960s/70s bold, geometric prints and matching chrome and white lamps, tables, and furniture. As a concentration of modernist, capitalist endeavors, Nosferatu's villa is the exact opposite of Count Dracula's castle. In Stoker's tale, the vampire's lair symbolizes the Gothic specter of the past, haunted by blood and memory and filled with piles of old, dead money. Here, the villa is the modern-day present looking forward, almost something of a non-place that facilitates the transport of its inhabitants into the future. The ghosts of Dracula's mansion have been replaced in this incarnation by adver-

---

103. The Illuminati were originally a Bavarian Enlightenment secret society founded in 1776 to fight religious superstition and the abuse of state power, but more recent fictitious examples often portray them as a dark, malevolent power controlling global wealth and power.

tising slogans, and its wolves by Fiat 500 bubble cars. Nosferatu is revealed to be the leader of a secret society of vampires that, when they are about to die, transfer their spirit into human babies who are then raised in the outside world, of which Alberto is one. He has therefore been watched from birth to ensure his safety so that he will one day take up his rightful place in vampire society—this is very similar to an idea used in the later Australian film *Thirst* (Hardy 1979), where the spirit of Elizabeth Bathory is reborn in the body of a human. This transference, or reincarnation, is represented as deeply unnatural within the film, almost demonic in a way and reminiscent of the eponymous child in *Rosemary's Baby* (Polanski 1968). At a time when artificial insemination was not yet commonplace, and indeed was problematic in Italy due to the influence of the Catholic Church (see Traina 1980), the film frames it as one of the many ways in which the bourgeois vampires control the world around them. Consequently, Nosferatu's correlation to industrialization and capitalism constructs the process as oppositional to environmental ecology and as another way to exploit the world.

The theme of industrialization and control of the natural world is introduced at the beginning of the film, with Alberto's life centered on the city and his job in the offices of the car factory. His urbanized life is disrupted by his trip out to Nosferatu's villa in the mountains, but the mountains are configured as the remit of the wealthy and off limits to normal city dwellers—not unlike Harker's journey to Transylvania in *Dracula*, with the young professional becoming increasingly disoriented the further he distances himself from civilization. As Alberto approaches the mountains, they become "an autumnal, foggy place, all gravestones and creepy villagers, including a mute wall-eyed geezer and a feral girl dressed in a fur coat and not much else" (Lines 2015). This cloaked, Gothic space of the natural environment between the city and the vampire's lair manifests a girl, Laura (Francesca Modigliani), to save Alberto from himself.

Almost inevitably, Alberto is at his most vulnerable when his car runs out of fuel and he is forced to stop at the village, a place utterly outside the purview of modern capitalism and where the houses/huts are built from stone. The girl is almost Eve-like when she appears semi-naked and seemingly from nowhere. She is visually connected to nature through her earthy and autumn colorings, seeing her with sun-tanned skin, brown hair and brown trousers and then joining Alberto on his walk through the countryside to Nosferatu's villa. Laura seems uninhibited, with little to no regard for possessions, offering Alberto a chance for a "natural" love and a stronger connection to the world beyond (without) capitalism. Laura and Alberto's relationship is then contrasted with Alberto's new relationship with one of the vampire's "brides" inside the walled villa. As Alberto abandons Laura outside the gates to the villa, he is escorted by Nosferatu's guard dogs in the form of two Fiat 500s,

Giuliano Esperati as Alberto Valle is surrounded by blood hounds in *They Have Changed Their Face*. Directed by Corrado Farina (Garligiano, 1971).

part of a larger group that are also seen to regularly patrol the grounds and chase down strangers.[104]

This seems to complete Alberto's transition between worlds, and as he enters Nosferatu's home (lair), he is inevitably seduced by the vampire's secretary, Corinna (Geraldine Hooper). She is purposely contrasted to Laura in having a very "unnatural" complexion: she is very tall, thin, sickly pale and androgynous in appearance, akin to a fashion model who is extremely stylized in dress and pose. While Laura is symbolically organic and connected to nature, Corinna clearly links to the stylized modernist furniture in the villa. Not unlike Archie in *Blood Car*, Alberto's denial of ecology and his descent into full-blown capitalism are signaled by his rejecting the "natural" girl in favor of the girl who fashions herself as a product for consumption. At the end of the film, Alberto, after killing Nosferatu, runs from the villa to go to Laura, who is still waiting outside the gates for him. However, she is now dressed smartly, knowing it is too late and seeing him for who he really is: an eternal consumer. She thus leaves him in the car and walks away, slowly disappearing back into the mists from whence she came. Meanwhile, Corinna

---

104. The idea of industry being vampiric and producing vampiric cars to consume society is also picked up in the slightly later Czech film *Upir z Feratu* [Ferat Vampire] (Herz 1982) in which Ferat is the name of the mysterious foreign car manufacturer taking control of the motor industry across Europe.

Chapter 4. The End of the End

has arrived at the gates of the villa, slowly opening them to invite Alberto back into his true home.

The true nature of Alberto's "home" is revealed just prior to his shooting Nosferatu in a business meeting at the Villa that Alberto eavesdrops on. The meeting includes philosophers, artists, priests, and businesspeople who openly talk of Hitler and Mussolini as inspirational figures and declare that "advertising is truth." As the meeting gets underway it becomes clear that Nosferatu's corporation not only owns cars but spans the entirety of consumerist production and marketing. It seems that these meetings occur on a regular basis, and the present gathering is addressing weekly production in the chemical section. They discuss chemicals, contraceptives, and bioweapons, among other areas of expertise, all of which are doing well except for detergents, whose sales have dropped due to their damaging effects on the water supply and little children. Although the vampires are working on a bio-friendly replacement, Nosferatu steps in to say that this would mean dumping the huge amounts of product they have already made, and so they decide to repackage the dangerous substance, rebrand it, and sell it anyway through their own chain of supermarkets. As the meeting continues, one of the department heads is reprimanded for allowing sexual relations between his workers, which negatively affects productivity and tarnishes the company's reputation (the vampires are curiously anti-reproduction), and he is subsequently expelled from the villa as punishment. He is then hunted and killed by the Fiat 500s, whose drivers then dump the body in a ditch on the villa's grounds.

Over the course of the meeting, Nosferatu Industries is revealed to have bribed a government official to legalize hallucinogenic drugs and has produced individual and family-sized packages of LSD in aerosol cans, which are already being distributed for free to poor families. An experimental filmmaker has produced an equally absurd and sadistic advert selling the product, which gets the approval of all members present, including the representative of the church. The drug, apart from being addictive and ensuring repeat sales, will also allow the vampires to control and influence its customers so that they buy everything they are told to from Nosferatu's many brand names.

Alberto has witnessed all this from a balcony that is out of view from the meeting room, and he is horrified by this display of total disregard for human life and the planet in favor of ever-increasing profit, so he makes to leave the villa. However, he is trapped in the building and confronted by Nosferatu, who dares his protégé to kill him. Alberto shoots him with a small handgun and invites Corinna to leave with him, but she refuses. He leaves the villa, and this is when he meets Laura and is rebuffed by her as described above—though an earlier scene also suggests she has been "reprogrammed" by Nosferatu's henchmen. His acceptance of Corinna as she opens the gates to the villa for him also signals his innate incompatibility with the natural world around him. As he drives his own car into the compound—his car here is actually representative of his former identity, and it has spent almost the entirety of the film parked outside Nosferatu's villa, until this closing scene[105]—he is followed by a swarm of the little Fiats until he parks outside of the villa. Once there, he gets out of the car to find Nosferatu alive and well—since he is a vampire, bullets have little lasting effect upon him—standing at the top of the stairs entering the building, waiting to finally greet him into his new, totally unnatural, vampiric life.

They Have Changed Their Face talks of the innate, vampiric corruption at the heart of human society that will guide it to inevitably consume itself and the planet for the sake of wealth. The next film is strangely similar though it envisions a future where this system is kept in perpetual balance, offering a kind of immortality to those who follow its ideology unquestioningly.

Snowpiercer is another film that is set in the near-future, 2031, though a post-apocalyptic one created by actions taken 17 years previously (i.e., the year the film came out). In 2014, in a desperate measure to reverse the effects of global warming, it is agreed to artificially change the climate by seeding the upper atmosphere with chemicals. Unsurprisingly, this goes dramatically wrong, causing a sudden and cataclysmic ice age—the suggestion being that fighting a chemically induced "fire" with more chemical "fire" will inevitably go wrong and trigger the opposite effect, a super freeze.[106] All life on Earth is killed except the humans, flora, and fauna that manage to board an ark-like circumnavigational train designed by transport magnate Wilford (Ed Harris). The train is huge and divided into many compartments providing sustenance, living space and entertainment for all the passengers, though these differ greatly in quality, depending on the ticket that was originally purchased, with the lower-class, economy carriages near the rear and the upper- or first-class carriages near the front. The worst lot is reserved for those in the tail section, who boarded for free but are kept in deplorable, prison-like conditions. The passengers in the tail section discharge certain duties on the train, but more importantly, young children are taken from the tail section at regular intervals by one of Wilford's representatives for some mysterious purpose. The ill-treatment that they receive, together with their terrible living conditions, has led the occupants of the tail section to stage various unsuccessful revolutions, and the narrative of the film revolves around the latest revolution,

---

105. This sees Alberto's story as a spiritual one that is finally consummated when it is reunited with his physical body, the car, and taken into the villa's grounds.

106. As with *I Am Legend*, this apocalyptic event for humanity is one that is embraced by the Earth, and even though it suggests the extinction of much of the life that formerly inhabited it, enough will survive—a point made at the close of the story when a polar bear is spotted in the distance.

Chapter 4. The End of the End

which is led by a man named Curtis (Chris Evans) who still remembers the horrors of boarding the train.

The train's hierarchy is shored up by religious and capitalist considerations alike, as explained by Wilford's representative, Mason, who tries to curtail the first stirrings of Curtis' insurgence:

> Order is the barrier that holds back the flood of death. We must all of us on this train of life remain in our allotted station. We must each of us occupy our preordained particular position.... In the beginning, order was proscribed by your ticket: First Class, Economy, and freeloaders like you. Eternal order is prescribed by the sacred engine: all things flow from the sacred engine, all things in their place, all passengers in their section, all water flowing. all heat rising, pays homage to the sacred engine, in its own particular preordained position [Bong 2013].

This makes for obvious comparisons to 21st-century America, which combines many features of neoliberalism with Protestantism and constructs a worldview where human worth is seen only in terms of financial returns as ordained from above. Wilford is thereby perceived as, if not God himself, the Father of the church of the train or, more specifically, the sacred engine. The "train of humanity" becomes both a microcosm, or mirror, of contemporary ideology (Radovic 2017, 65) and a means to keep that same humanity separate from the world around them, breaking the bonds forged by "a billion deaths" (Spielberg 2005) and detaching humans from their own ecosystem.

The church of the train constructs hierarchies based on wealth—though this seems a curious distinction in a world where money no longer has any real meaning. Consequently, onboard status is an originating position, almost as if one was born into it, that remains unchanged from the moment one enters the train (i.e., if you purchased a first-class ticket, you remain on the first-class carriage forever). The train itself thereby becomes something of a synecdoche of the now-lost capitalist world, a seed carrying the genetic code, as it were, fully invested in maintaining the same class divisions and prejudices so that it can sprout and regrow once the Earth is made habitable again. The symbolic role of actual money is therefore filled on the train by blood and blood sacrifice as a fundamental part of the onboard economy and a basic ingredient of all the exchanges that take place on board. This explains the self-sustaining nature of the train and its population and why the planet makes sure that it is never allowed to meaningfully affect the wider environment.

The economy of the train cannot contain and produce enough food and water to keep all its passengers alive, suggesting that cannibalism of some sort is being conducted, with or without everyone's knowledge. The passengers in the tail section are inevitably the first to be eaten, as they are the most disenfranchised and overpacked compartment. Indeed, Curtis' story of the first weeks and months on the train reveals as much. He tells of how those who were allowed to board for free were rounded up by Wilford's troops, and a thousand of them were forced into the large carriage without food or water for over a month so that only the strongest survived, and only then by eating the weak. After being "born" into the train hierarchy through consuming flesh, they have since been kept alive on a constant supply of protein bars, which immediately conjure up thoughts of the film *Soylent Green* (Fliescher 1973) in which human flesh, unbeknownst to the public, is used as a food replacement. *Snowpiercer* tries to avoid such a conclusion, at one point showing huge cockroaches as the main ingredient of the bars, but inevitably the narrative keeps returning to it.[107]

The suggestion of cannibalism is introduced in one of the early scenes when Wilford's representative, assisted by soldiers, arrives to fetch a child from the tail section. As the representative departs, an inmate attacks her with an improvised projectile, leaving a cut on her forehead. The representative wipes her finger over the wound, gathering up the blood and licking it off. A similar image recurs after the uprising has begun and Curtis and his followers move through the train carriages to what they believe to be the water processing carriage but instead find a compartment full of masked men in leather aprons holding axes, who are promptly ordered by Mason to kill all the insurgents. These masked men closely resemble butchers, playing off of the idea of humans as meat and a product to be dissected and packaged for consumption. The ensuing battle is centered around large sprays, spills, and pools of blood—a dramatically choreographed sacrifice of humanity for the continuance of the ideological ark of the train. Curtis, while sacrificing his friend, captures Mason to use as a human bargaining chip to reach the sacred engine at the head of the train. Thus, all actions require the spilling, and possible consuming, of blood, which never satisfies the train but only exacerbates its insatiable hunger.

Curtis and the insurgents pass through a variety of carriages on their way to the front: one full of plants, another constructed as a walk-through marine tank full of sea fish, and a third a schoolroom where the children of the first-class passengers are inducted into the ideology of the train (these children are never taken by Wilford, it should be noted). Curtis finally reaches the engine carriage and is invited in to speak to Wilford, who reveals that all the previous revolutions, including the current one, were purposely orchestrated to reduce the population in the tail section and maintain the pre-ordained balance on the train. While delivering this explanation, Wilford is eating a large steak, which he also offers to Curtis.

---

107. The film never explains how it would keep and farm such a large number of insects, never mind what they would be fed on if not in some measure the dead bodies of those dying in the tail section.

Chapter 4. The End of the End

Chris Evans as Curtis (center) is glamoured by the engine in the presence of Ed Harris as Wilford, its creator and caretaker, in *Snowpiercer*. Directed by Bong Jon-ho (Radius-TWC, 2013).

The steak raises several questions, as we have not seen any kind of livestock on the train—and steak would not have survived 17 years in a freezer if such a thing existed on the train—other than the human livestock kept in the tail section. As Wilford continues to glamour Curtis with the inherently cannibalistic ideology of the engine, he offers him the chance to take over his job, but just as the younger man begins to fall under Wilford's spell, Yona (Ko Asung) runs in and lifts one of the floorboards in the engine, revealing a young child encased in parts of the train's mechanism.

The child reveals that the sole purpose of the passengers in the tail section is to provide young children of exactly the right size to become parts of the engine. The precise details of the children's work in the engine are never disclosed, and it remains unclear whether they move things manually to keep the various gears and pistons in working order or whether the children are actually human batteries, but either way, they never come out of the engine alive—in solely requiring children, the engine literally consumes the future of humanity to continue its journey and maintain its consumerist world. Curtis sees the missing child, Timmy, put into the engine as a "new part" and sacrifices his own arm to stop the gears long enough to get the boy out. This sacrifice enables Curtis to atone for his part in eating fellow passengers in his first months on the train, when the tail section literally consumed itself to survive. This last sacrifice of his arm seems to break the cycle of ideological vampirism embodied by the train. Another one of the last surviving rebels, Namgoong (Kang-ho Song), sets off an explosion that dovetails Curtis's sacrifice, bursting a hole in the walls of the engine, a "symbolic 'rupture'—blowing up the system" (Radovic 2017, 67). The train is derailed by the shockwaves and sent careening off into the surrounding snow drifts, where only Yona and Timmy crawl out into the snow and start walking away from the train. As they look into the distance, they see a polar bear, signifying that the ice age has begun to recede and that now, as the last remaining seed of capitalism is dead, the environment is ready to embrace humanity once again.

Snowpiercer can also be read as an example of vampiric technology used to sustain humanity and, in a way, help to keep it in balance with the wider ecology around it. In this sense, it is the engine itself, as a product of a vampiric ideology, that configures the vampire—the undead creature that consumes human blood to survive. However, not all vampiric technology requires blood, or even the destruction of humanity, to realize some form of ecological balance.

John Carter, Andrew Stanton, 2012 / *Dracula*, Cole Haddon, 2013–14

John Carter is not an obvious vampire film. Indeed, the original 1912 novel by Edgar Rice Burroughs, *A Princess of Mars*, of which the film is a very free adaptation, does not contain any vampires at all. However, the movie by Stanton necessarily condenses much of the plot and turns one of the novel's less-dynamic groups of characters into the movie's arch-villains. The Therns, who in the book are something of a Martian Holy Order gifted with a lifespan of 1,000 years, are immortal, interplanetary vampires in the film.

The movie is set in 1868 and tells of an American Civil War veteran, Captain John Carter (Taylor Kitsch), who is transported to Mars (Barsoom) and becomes the savior of the dying planet. Barsoom is a dried-up world inhabited by green Martians named Tharks and red Martians who are divided into two groups centered around the cities of Helium and Zodanga and who have been at war for a thousand years.[108] The Therns, unlike in the book, are not native to Barsoom but are known as part of the planet's creation story as servants to the goddess Issus, almost mythic in status, as their last recorded appearance occurred in the distant past. The Therns are the only ones who know of the existence of "the ninth ray," a source of almost unlimited energy hidden within light, harnessed by the original Barsoomians and utilized through specialized technology. The Therns have ensured that anyone who knew of the existence of this energy is dead, save themselves, so that they can use it for their own nefarious ends. This secrecy extends to the one remaining shrine to the goddess along the sacred river, the Iss, which is constructed of

---

108. The narrative contains many problematic representations of race and Hollywood's ubiquitous need for a "white savior."

an inert fiber technology waiting to be re-energized by the ninth ray. The novel represents the shrine, and similar constructions, as atmosphere creators that would restore the planet's ecology if properly used, but the Therns of the film have re-appropriated this technology for their own use. In fact, the Therns are so secretive about the ninth ray that they have also made their own existence a myth as well, only appearing to those whom they can manipulate. They are further aided in their nefarious efforts by their ability to change their appearance at will—at one point in the film we see the leader of the Therns, Matai Shang (Mark Strong), talking to Carter while seamlessly transforming from a bodyguard to a chamber matron to an old woman to a Zodangen soldier and finally to a guard again. In their true form, the Therns appear to be exclusively male, white skinned, and bald, wearing long, gray robes. They also wear a large golden amulet with a gem at its center around their necks—the gem harnesses the power of the ninth ray and allows the wearer to change appearance and travel great distances at will.

The Therns' pale and bald appearance resonates with Murnau's Count Orlok and the many incarnations that followed of mysterious visitors who hide in the shadows, exerting their influence on the innocent and unwary from afar. *Nosferatu* follows this pattern in the visible effect on Ellen and Knock the moment the vampire leaves its home and starts making its way to Wisborg. *Dark City* (Proyas 1998), a film about vampiric parasites from outer space, presents a more modern example. The film features alien beings called The Strangers who are extremely pale, bald, male humanoids dressed entirely in black leather. They have constructed an artificial city that floats in space, and they have filled it with humans so that they might study aspects of the human brain. Every night, The Strangers manipulate the city and the memories of the humans within it so that they live completely different lives every day. The precise mechanism by which this manipulation sustains the alien parasites is unclear—whether they are energy, emotional, or psychic vampires in some way—but it is this continual manipulation of their hosts that keeps them alive. The Therns in *John Carter* are very similar. Not only are they immortal—as is Orlok—but they also seem to survive by drawing energy from their host (the chosen people on whatever planet they have moved to), manipulating and ultimately destroying them. As Matai Shang explains to John Carter:

> We are eternal, … history will follow the course we have set Earthman, and we have chosen Sab Than [the leader of Zodanga] to rule it. The ninth ray must remain in the hands of mindless brutes that we can control … we've been playing this game since before the birth of this planet and will continue to do so long after the death of yours. We don't cause the destruction of the world captain Carter, we simply manage it. Feed off it if you like. On every host planet it always plays out exactly the same. Populations rise, societies divide and wars spread and all the while neglected planets simply fade [Stanton 2012].

The Therns thereby feed upon human conflict/emotions and use their powers to create such conflict by manipulating a planet's inhabitants to fight each other until they have destroyed both themselves and their home world—apparently the war between Helium and Zodanga had already been raging for 1,000 years when Carter arrives.[109] In this sense, the Therns embody the idea of consumerism itself, not unlike the engine in *Snowpiercer* that thrives upon humans with little regard for the wider ecology, indeed suggesting that in its essence it is the negation of life, or death as embodied in the mindless desire to consume everything, including themselves. This vision of consumerism-as-death also emerges in *Jupiter Ascending* (the Wachowskis 2015), where the vampire Abrasax family, who unlike the Therns are immortal and stay young by processing entire planets into an elixir in which they bathe, manifest a similar stark disregard for the eco-environment in the name of consumerism and their own personal satisfaction. The Abrasaxes are extremely conspicuous in their consumption, living lives of god-like opulence and inspiring awe in those around them. The Therns spurn such a life in the spotlight, however, and although sunlight appears to have no effect on them, they remain hidden as an omniscient, supernatural presence, like Orlok, rather than act as the brazen devourers of worlds.

Another difference between these two kinds of all-consuming space-vampires emerges in their use of technology and its relation to the environment. The Abrasax family uses science and technology to manipulate and control the natural environment, which is most clearly seen in the film by the large amount of hybrid creatures that they have genetically spliced together to serve them: humans crossed with bees and wolves, or other alien species interbred or mutated to produce armies of loyal soldiers, trackers, or aggressive henchmen. The Therns, in contrast, create technology by manipulating a natural, if elusive, energy source that seems to enliven the devices through which it is channeled. They are thus more closely aligned to the environment and able to harness its power to their own ends to create an organic kind of vitalized architecture—a living network or arterial system of energy that can power a planet and create and potentialize life, as exemplified in the sacred temple on the river Iss.

Notwithstanding that the Therns use this power to promote their own interests, it can also be seen as something of a universal regulator. As soon as a lifeform has evolved to the stage of being able to negatively affect its own home world and potentially planets beyond that, the Therns are then drawn to them to protect the rest of the solar system—a kind of intergalactic regu-

---

109. The fact that the Therns have replicated this scenario on many planets suggests it's not just humans they prey upon but any lifeform able to produce similar kinds of emotional/psychic energies.

lator, as seen in *The Day the Earth Stood Still* (Wise 1951), for example. This would also explain their growing interest in the Earth, in part drawn there by the American Civil War and developing technologies at the end of the 19th century that mark out humanity as the next species that could quickly become a danger not only to their own world but to the inhabitants of other planets as well.

John Carter's actions on Mars are shown as a means of possible reparation between the planet and its inhabitants, though the Therns are not dead, and like Dracula, time is inevitably on their side. This idea of harnessing natural energy for the ultimate good of the planet is taken up in the next film, which is more earthbound than the previous examples.

Cole Haddon's *Dracula* is a curious series that blends elements from the history of Vlad the Impaler with a steampunk version of Count Dracula, though not so much as a threat to civilization as its potential eco-savior. The titular vampire of this film is trying to save the world from itself, or rather, its consumerist impulses, which are knowingly fueled and fulfilled by a nefarious society that is more vampiric than Dracula himself. Most of the action takes place in an alternate late-Victorian London, but the forces at play in that world were set in place hundreds of years beforehand in the 15th century. The backstory presents the historical Vlad II, ruler of Wallachia, falling foul of the Order of the Dragon, who then turn him into a vampire and bury him "alive" in an unmarked tomb—they use a witch to perform the transformation from human to undead. The Order also has some historical truth and saw its mission as protecting Christendom from the invading Ottoman hordes; in Haddon's series, they have far more selfish and far-reaching intentions to amass power and wealth and basically indoctrinate the "free" world into supporting their aims.

As part of this, they are intimately linked to Western colonial endeavors, particularly those of Great Britain, seeing them reflecting other real-life colonial "societies" such as the East India Company.[110] The historical East India Company was founded in 1600 and received a royal charter to effectively monopolize all trade between Britain and East Asia. Identified as the world's first, and most powerful, corporation (see Roy 2012 and Robins 2012), it aggressively enforced capitalist ideology across the globe, wantonly commodifying lands, people, and resources for economic gain. Haddon's Order connects such consumerist colonialism with other darker powers so that while it seems inherently opposed to supernatural entities, such as vampires, it freely employs psychics and vampire slayers to rid itself of such nuisances. In fact, the series builds both groups, the Order and vampires, as oppositional

---

110. There is also something of the Freemasons mixed in here, seeing it as an extremely secretive society that looks after its own, particularly in terms of business and accruing wealth.

in some sense, and Dracula crystallizes the nature of this irreconcilable division.

Dracula first appears as a desiccated, undead mummy entombed deep beneath the ground, subsequently unearthed by Doctor Van Helsing. Digging down into the ground to discover the tomb, Van Helsing kills his guide, allowing the blood to fall into the mouth of the mummified Vlad, with the life-giving fluid slowly bringing him back to life.[111] Dracula is thereby positioned as intimately linked to nature, emerging reborn from the earth. He seems curiously in sync with the environment around him and only takes from it what he needs to survive. This is in contrast to the Order, who seem driven to consume the planet's fossil fuels in a way that is against the good of humanity and the planet in terms of its damage to the land and its peoples—who are willfully plundered—exacerbated by the emissions from the use of fossil fuels that have wider, long-term ramifications for environmental stability.

This free reworking of historical detail adds to the "steampunk" construction of the Victorian Britain that Dracula enters and which the Order seems to control, presenting it as a historical mise-en-scène that is sharply contemporary in its environmental and ideological intent. Consequently, the 19th-century setting talks of 21st-century environmental concerns around climate change and the dire ramifications of the refusal to find alternatives to fossil fuels such as oil and coal. Indeed, it is oil that provides the focus of the Order's ongoing colonial, consumerist project, as its representatives try to convince the British government to dispatch forces to the Middle East to secure the oil-rich lands there. This late–19th-century enactment of the historical Order's intent of beginning a crusade to invade Ottoman territory is one that also foreshadows the breakup of the Ottoman Empire at the end of World War I, largely so that Western powers could gain control of the region's oil reserves. Such a Manichean approach to the environment is also apparent in the differing views on technology of the two parties, where the Order is rooted in military might—as a technology of colonialism/consumerism—but also a backward-looking industrialization that looks not to new technologies but to a forced dependence on what is already in existence, as opposed to Dracula, and more especially Van Helsing, who see technology as an organic, futuristic, almost supernatural creation. This excess produces the same kinds of "magic" that Stoker's Van Helsing speaks of when noting, "Let me tell … there are things done to-day in electrical science which would have been deemed unholy by the very men who discovered electricity—who

---

111. There is a similar scene of vampiric unearthing, though in South America rather than Eastern Europe, in *Bordello of Blood* (Adler 1996), though here it is Lilith rather than Count Dracula who is brought back to life.

would themselves not so long before have been burned as wizards" (Stoker 1996, 205). Following on from this, although the series sets itself up as a traditional vampire narrative of slayers versus the undead, with the obligatory contemporary twist of the human "monsters" being the true villains of the piece, of more interest here is the way that the story is underpinned by the differing uses of, and approaches to, technology.

The technology of the Order is hidden in plain sight, enabled by the stark hierarchies between those who have access to electricity, motor cars, etc. and those who do not—though it should be noted that this vision of a Gothic Victorian London is rather sterilized in comparison to other similar, contemporaneous series, such as *Penny Dreadful* (Logan 2014–16) and *The Frankenstein Chronicles* (Langford 2015–present), which manage to create a far more menacing environment. To highlight this idea of hiding in plain sight, Dracula, under his false identity of American industrialist Alexander Grayson, builds his revolutionary powerplant with the aid of Abraham Van Helsing in one of the poorest areas of London to more easily provide free electricity to those in the near vicinity and upset the previous consumerist hierarchies.

Van Helsing also suffered huge personal losses inflicted by the Order—his wife and children were murdered by them—and rather than just seek the more traditional means of revenge, such as killing those who were involved, he has decided to try to destroy the secret society completely by cutting off their means of economical control. He thus hatches a plan to generate large amounts of electricity and provide it for free across London via a revolutionary WiFi service—this idea was actually thought of by the inventor Nikola Tesla at the start of the 20th century but was never realized (see Laskow 2018). This plan leads to the construction of a factory in London that is overseen by Grayson and stylized as pure steampunk Victoriana—it resembles the engine room of a large seagoing vessel. Exactly what it uses to produce this energy is never fully explained, and although there is a hint that it might be nuclear, it certainly bears no relation to 20th- or 21st-century nuclear powerplants. Curiously, when the generator explodes, it does not emit radiation or similarly poisonous or deadly emissions, suggesting that it operates on excessive life energy. Indeed, the source seems almost supernatural in nature, creating a curious, glowing orange light in the large canisters above the generators. This impression is reinforced by the first public demonstration of the generator when Grayson gathers together a large group of London society to his "coming out" ball and asks all his guests to hold electric lightbulbs in their hands, which magically begin to glow brighter and brighter as the powerplant begins to generate energy. Unlike the other sources of power available in this fictional London, Grayson's powerplant seems to use no oil nor coal nor indeed any fuel at all and equally creates no waste or pollution: it seems to come from the air and return to it. In a commentary on 21st-century resistance to environmentally friendly power sources, particularly by the established and fossil fuel–reliant corporations, the Order does all it can to discredit Grayson and destroy the plant.

At the end of season one—the series was not renewed for a second—the Order, on the cusp of losing its battle against Grayson, manages to plant a bomb in the factory and destroy it. Van Helsing is enraged not only by their act of sabotage but also by Grayson's misdirected attention. Grayson discovers someone whom he believes to be the reincarnation of his dead wife and tries to become human again to be with her. Van Helsing, realizing he will not be able to surprise his enemies in such a way again, turns the children of the Order's leader into vampires, allowing them to consume their own father in a somewhat ironic take on the way the Order consumes the planet that gave it life. The chance of a sustainable world—even in this otherworldly, futuristic, steampunk 19th century—where humanity can continue its capitalist lifestyle without harming the environment has gone forever. Van Helsing, who was a champion of economical and ecological change, has resorted to straightforward violence to exact his revenge on all those who have wronged him.

Haddon's series begins to describe a world where the environment can be saved by technology rather than harmed by it. A more extreme version of this would see a world where vampiric technology destroys the remnants of humanity so that it can begin a symbiotic evolution with the environment itself. It is this vision of an increasingly non-human future and a reinvention of the ideology of consumerism itself that comes center stage in the last two films in this chapter.

Ex Machina, Alex Garland, 2014 / *Autómata*, Gabe Ibáñez, 2014

Garland's film is not obviously one about vampires, but there is much within it related to the vampire genre. Equally it begins to describe the vampiric nature of technology and the post-(consumerist)human. As I have written elsewhere (see Bacon 2018), the film follows the plot of the opening chapters of Stoker's *Dracula*: a young professional is invited to meet a mysterious man in his lair in the "land beyond the forest," where he is seduced and ultimately left for dead as the vampire leaves its home and travels into the heart of civilization and the "whirl and rush of humanity" (Stoker 1996, 22). *Ex Machina* establishes its vampiric credentials by following this same story arc, with a young computer programmer, Caleb (Domhnall Gleason), invited to the "lair" of his company's CEO, Nathan (Oscar Isaac), which can only be reached by helicopter. Caleb has been invited to help Nathan test his latest humanoid AI, Ava (Alicia Vikander), but as the film unfolds, he is seduced by Ava while facilitating her escape from the compound. The film ends with Nathan dead and Caleb locked inside the complex while Ava escapes by helicopter—it was due to return Caleb back to civilization—where she blends into the mass and whirl of humanity. The twist in this film is that the vampire is in fact Ava and not Nathan and that she manifests the shapeshifting, immortal qualities of Count Dracula.

Indeed, it is the mutability of Ava that aligns her with the idea of the vampire-cyborg as described by Allucquére Rosanne Stone via Donna Haraway (2013), which sees a blending and hybridizing of the human and the technological, creating new subject positions beyond gender and other heteronormative categorizations (Stone 1995, 178–83). In many ways, this is a vampirism in all characteristics except the most culturally obvious, that of blood-sucking, for the cyborg is immortal, shapeshifting, and beyond all cultural norms. Though as seen in *Ex Machina* and *Autómata*, the vampire does not need blood to sustain itself but can also exist through energy transference and bodily hybridization, as in the grafting and reconfiguring of parts to produce something of a self-supporting evolution that takes nothing from the ecosystem around it. Garland's film begins to sketch out just how that might work.

When we first see Ava, she seems just like one of Count Dracula's brides, a seductive form largely of lips, eyes, and curvature that takes substance almost out of the very atmosphere of the vertiginous, reflective corridors of Nathan's underground lair—in a curious inversion of Dracula's castle that stands on a mountaintop surveying the land around it, here the lair is underground, hidden from view, yet both interiors create a similar, otherworldly space where the normal rules do not apply. In fact, all the humanoid AIs built by Nathan are seductive, beautiful women in appearance, intimating their categorization as property to be bought and sold. This commodification of human intelligence in general and females in particular is showcased by the designation of Ava and the other AI in the lair, Kyoto, as fully functioning sexbots, and even though they are the most sophisticated and advanced androids in existence, they are treated as little more than sexual playthings by their inventor.[112] Ava's designated function informs her appearance—she is made of a naked metal skeleton but has more solid structure around the areas of her hips and breasts, with imitation skin over her hands and face. She is thus endowed with enough of a human appearance to appeal to the more

---

112. There is something of a Marxist reading of this situation in that the rich/ruling classes purposely keep their workers/subjects enslaved and captive for fear of what might happen if they ever gain their freedom, something that is intimated in *Ex Machina* by Nathan's description of his invention as having the potential of a nuclear explosion if released into the world.

impressionable man, namely Caleb, as a human woman trying to escape a technological body. Ava's appearance is engineered to appeal to Caleb, and, arguably, her consciousness has been linked to the internet, where it was able to monitor the young programmer's web browser and find out his preferences on porn sites.

This premise highlights the way in which corporations sexualize their products to more effectively target consumers and boost sales. It equally highlights the vampiric nature of Ava herself, who is in fact a genderless machine that takes on different shapes and forms to achieve its goals—not unlike Stoker's Count Dracula.[113] This point is brought home at the end of the film just after Ava, with the help of Kyoto, has killed Nathan but was damaged in the struggle with him and so needs to repair herself before leaving the lair. This act of self-repair is of particular note here and has an even more essential role in the last film in this chapter, as it implies a system that does not require human interference and is also a synecdoche of the ecosystem's own self-regulating and restorative processes. Ava then goes to a storeroom where Nathan keeps all the disassembled parts of her predecessors, a cross between a charnel house and a literal representation of Bluebeard's Castle.[114] In the kind of vampiric performativity that Haraway and Stone describe, Ava chooses replacement limbs, large areas of skin and even long brown hair to construct a version of herself that will blend easily into the outside world—one assumes she has gained this image through her earlier internet searches.

There is much within this that intimates consumer culture in the 21st century, where one can transform oneself via cosmetic surgery and other methods to look like someone else.[115] Ava embodies the perfect consumer who chooses their looks and clothes from the internet, an idea that is carried on in the scenes as she leaves the lair. As she—and Ava has purposely taken on the persona/appearance of a woman—walks into the sunshine for the first time, she could be a fashion model in a commercial selling both her clothes and a lifestyle as she, wearing a white lacework dress, wanders through the luxury living room area of the lair and out into the brilliant sunshine to board a helicopter to be taken to the city. Hers in not a blind consumerism but a form of mimicry that is an act of resistance, one that could lead to the end of unchecked human consumption and a return to ecological balance once more.

---

113. It is not made explicitly clear in the narrative just how aware Nathan is of what Ava achieved while connected to the internet, though the lair is purposely shielded from WiFi signals, etc.

114. The tale is of Bluebeard bringing his new wife back to his castle, only for her to find the dead bodies of all his previous wives behind various doors within it.

115. While there is little explicit inference that Ava gains any power by doing this, other than influencing the gaze of the outside world, power sources are never mentioned during the narrative, suggesting there is some form of internalized energizing system. As such, she might have gained some kind of dynamism from the process.

Chapter 4. The End of the End

At the close of the film, Ava has not suggested that she will set about the downfall of humankind, as Steffen Hantke has pointed out in his reading of the film (see Hantke 2019, 111–18). Indeed, she does her best to disappear into the crowd, to be lost from our view, and, one assumes, from surveillance cameras as well. This also resonates with Stoker's vampire, who does his best to vanish in the hustle and bustle of civilization to better attack its unsuspecting citizens. Before his death, Nathan suggests the danger of his invention if it was set free in the world, and there are enough hints in the earlier parts of the narrative that she might not be overly fond of humanity and may have the skill to replicate herself, and she appears to have no protocols in place that ensure the sanctity of human life as a default system in her memory processors.

The world she vanishes into is obviously a future version of this one yet not so far in the distance as to be unrecognizable. The earlier stages of the narrative construct humanity as still being unaware of, or unwilling to change, its effects on the wider world, requiring that the ecosystem itself needs to act in its own defense. In this sense, Ava begins to recall other narratives under discussion in this study where the vampiric plague emerges from the heart of the landscape, as she too is "born" from the soil of an uninhabited land in the middle of a forest, with Caleb fulfilling the role of the unsuspecting (naïve) Western traveler who releases it into the world. Ava, while being a mechanical/technological creation, can be seen to be "born" from nature and the wider ecosystem to stem the uncontrolled spread of humanity and its wanton consumption of the planet's resources. She is then the new "Mother" (Eve) of the next Creation, one that no longer places humanity at its center. This continues into the last film in this chapter, which more clearly intimates what the future of the Earth without human consumption might look like.

Autómata is a very different film from *Ex Machina* that more explicitly reveals the nature of the cyborg-vampire and its final independence from humanity. Not unlike *Snowpiercer*, it takes place about 30 years in the future from the time of the film's broadcast, but rather than a world under the thrall of a new ice age, it presents a world in the process of becoming a never-ending desert. In the 2030s, solar flares caused irrevocable damage to the Earth's environment, devastating large swaths of the natural ecosystem as well as water supplies—though there is more than a hint that humankind either had a hand in this catastrophic event or exacerbated its effects. As a result, 99 percent of the world's population was killed, with the remaining humans grouping together in walled cities to try and protect themselves from the encroaching sands. The survivors have built hordes of robots, called Pilgrims, in an effort to save themselves from extinction. The Pilgrims perform all manner of tasks to support the survivors and generally try to keep them from succumbing to the increasingly harsh conditions of life. Unlike Ava, these robots are quite basic in appearance with featureless, dish-like, plain faces with two red lights for eyes. They were purposely built with two protocols to protect their human makers and to restrict their evolution: first, they cannot harm any form of life, and second, they may not repair, modify or alter themselves or other robots in any way.[116]

The story begins in 2044 at a time when the robots that were once seen as the saviors of the human race are now viewed as a nuisance, mainly because of their inability to stem the flow of the encroaching desert. In this apocalyptic vision of the future, the surviving humans insist on trying to maintain a consumerist lifestyle even in the face of inevitable extinction. Consequently, the city is full of tower blocks with a constant stream of huge holographic adverts playing in the sky—the mise-en-scène is very much in the vein of *Blade Runner* (Scott 1982)—most of which seem to feature gyrating, naked women, producing an atmosphere of sexualized consumerism not unlike the world intimated in *Ex Machina*. The new society is still obsessed with money and material wealth rather than with securing its own survival, as evident in the central theme of insurance payouts that drives the story itself. The Pilgrims are extremely expensive machines, so any damage to them would bankrupt any ordinary citizen, and similarly, any faults to the robots would force the company, ROC, to pay out huge amounts of money. The plot revolves around a suspected fault in a Pilgrim that leads to a detective trying to find the culprit responsible. The urgency in the case highlights the extent to which this society is sharply divided between the very few ultra-rich and the majority who live in squalor, creating a dystopian vision of exhaustion and despair where life is still a commodity to be bought and sold, even though money will soon become worthless.

As noted earlier, the Pilgrims are at the bottom of this hierarchy. They are used for manual labor, and when they no longer have an owner they are forced to "live" in rundown areas of the metropolis or in makeshift shanty towns just beyond its walls. Not unlike in *Daybreakers* or *Snowpiercer*, the society is inherently vampiric in the sense that the wealthy "feed" on those at the bottom but also in the way that many of its inhabitants are more than human and performatively vampires, as they are augmented with robotic implants and replacement limbs taken directly from the Pilgrims. This more clearly represents the kinds of hybridity that Stone identifies as vampiric and

---

116. Such protocols are inevitably linked to the three created by the writer Isaac Asimov in 1950 to regulate robots in his *Foundation* series of books: 1. A robot may not injure a human being or, through inaction, allow a human being to come to harm; 2. A robot must obey any orders given to it by human beings, except where such orders would conflict with the First Law; and 3. A robot must protect its own existence as long as such protection does not conflict with the First or Second Law. This was subsequently amended/updated in the series with an extra protocol: A robot may not injure a human being or, through inaction, allow a human being to come to harm, except when required to do so in order to prevent greater harm to humanity itself.

extends to Pilgrims that have been specifically designed for the sex industry and more obviously blur the robot/human divide. Here, robots are given a more obvious female appearance, though not to the extent seen in Ava, so that they purposely bridge/blur the divide between technology and humanity, the living and dead. In fact, it is this divide and just how transformationally vampiric the Pilgrims are that underpin the narrative.

The story follows an insurance investigator, Jacq Vaucan (Antonio Banderas), checking on a report that a Pilgrim robot was found not only repairing itself but altering its own construction. This suggests that someone, a "clock-smith," has been tinkering with the Pilgrim and overwriting one of its founding protocols, which is believed to be impossible. The robot is shot in the head and "killed," but the autopsy reveals that it contains parts from several other robots—all parts have a robot-specific serial number lasered onto them—all still listed as operational. Jacq traces one of the parts to a specific robot and heads to its place of employment on the city wall, but it manages to escape into the wasteland beyond. Jacq tracks it to a container, but before he can react, the robot sets fire to itself—again an action that it should be unable to commit. Returning to the site of the original crime, Jacq discovers a hidden nuclear battery cell—a metal ball about 5 cm in diameter—which could power a machine indefinitely. In trying to trace this metal ball and understand its implications, Jacq draws the attention of the owners of ROC, who send killers to dispose of him and anyone else who is aware of the issue with the robots. The investigator escapes with a sexbot, Cleo, into the radioactive desert far outside the city—one of the causes of the desertification of the planet seems to be high levels of radioactivity in certain parts of the globe caused by the solar flares. They are joined by three other robots who then lead him to a cable car station on the edge of a canyon, where they meet the "clock-smith" who is in fact not a human but another robot.

The robot reveals that when ROC made their first robot it did not have any protocols installed within it and that eight days after it was built, it had already evolved into an intelligence beyond human understanding. The last action that its makers asked it to perform was to encrypt the protocols into a robot core that they could then place in all subsequent robots, operating on the assumption that only an evolved robot/AI could understand and break the code, as it was beyond human comprehension. ROC then destroyed the original robot. However, as noted in the film, "life will find a way,"[117] and even though nothing of the original robot exists, another one spontaneously goes beyond that encryption to change its own protocols, becoming more than just

---

117. Thanks to Hadas Elber-Aviram for pointing out that the same line is used in the film *Jurassic Park* (Spielberg 1993). Interestingly, both instances signify life instigated by humanity that quickly goes beyond human control.

a machine. Once the remaining "repaired" robots meet at the edge of the canyon, they proceed to make a new machine, a new kind of robot, one that can think for itself without protocols and is entirely beyond human interference.

The leader of the robot group explains to Jacq that they wish humans no harm, as mankind is a dying race anyway, and want nothing more than to begin their own life in the radioactive desert—the radiation seems to have no ill effects upon them. Jacq is moved to give them the nuclear power cell, allowing the robots to begin their own process of evolution without recourse to the city. The robot explains that humanity will live on in them, as they were born of human invention and designed to their scale and comprehension of the world—they are thereby on some level innately products of the Anthropocene, although the robots' ongoing existence will not affect the planet in the same way—though it fails to comment that the new robots will not be like that. The new robot is of interest, as its design does not come from human examples that have shown themselves hugely lacking in terms of resilience and adaptability but is from another natural source: insects. During its time at the cable car station, the robot leader has noticed the most persistent lifeform in this new world, the cockroach, and so uses this as the template for this new form of life that will be able to evolve and grow in the planet's harshest environment—a new start for a world on the edge of extinction.

The denouement of the film sees the hired killers of ROC dispatched, with their leader pushed to his death by the new kind of robot, proving it has no protocols to limit it other than its naturally developing sense of right and wrong, and it chooses not to kill Jacq even though the investigator is pointing a gun at it. Jacq's wife and newborn child were brought to the location as a way to make him turn himself in, but now they are the only humans alive at the canyon edge. Jacq helps to lower the robots down to the irradiated desert below before driving away to try and find what used to be the coast.

This ending suggests two very different beginnings: the robots are assured in their new lives, heading toward a reality where they know more will join them as the human population gets ever smaller; Jacq and his family head toward a dream—the image of a child playing in waves on the ocean shore occurs throughout the film in relation to the investigator, suggesting a connection/yearning for a return to ecological balance. The close of *Autómata* resonates with the end of *Ex Machina*, with artificial lifeforms using mimicry of their former masters as a weapon of resistance—seeing those things that were once considered possessions now claiming their own agency. Ava, like the Pilgrim robots, still carries the traces of her former human masters, but what happens beyond that is unknowable, and she, like the cockroach robot, is ultimately not restricted to humanoid forms and, in true vampiric nature, could even live as a separate consciousness online. *Autómata* especially suggests that human consumerism will inevitably destroy both mankind and the

Chapter 4. The End of the End 153

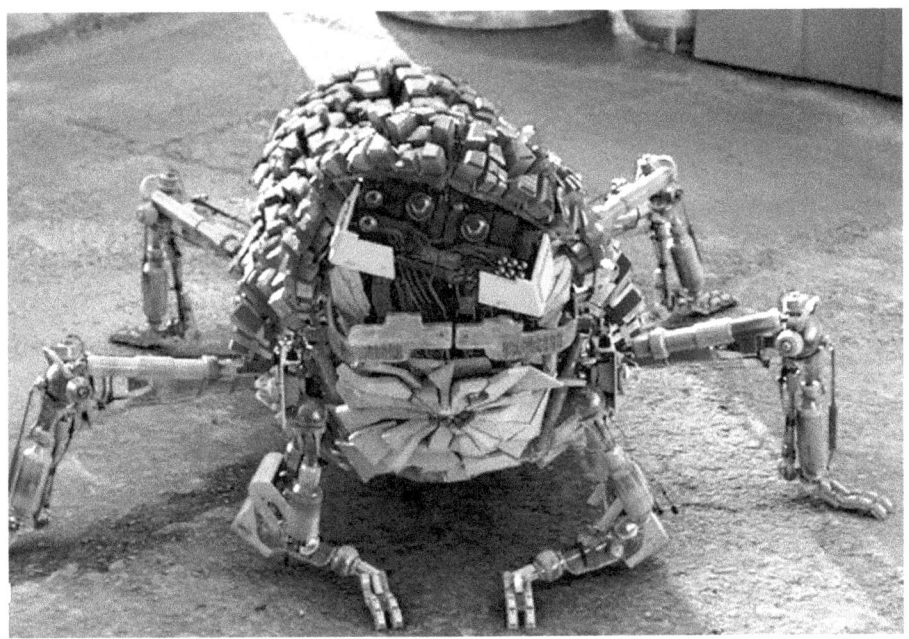

Evolutionary life in *Autómata*. Directed by Gabe Ibáñez (Contracorrientes Films, 2014).

planet itself, being a self-absorbing and self-consuming ideology that thinks of nothing outside of its own continuance. However, it further suggests that even after the most extreme of consequences, the planet's ecology will find some measure of balance, and life can emerge, develop and grow from the most unlikely of places. The solar flares can been seen as a broader, interplanetary regulator, as outside forces are prepared to step in and protect the wider solar system from the dangers of human consumerism and disregard of local, global or galactic ecosystems, a subject that forms the basis of the fifth and final chapter of this book.

Chapter 5

Vampire Ecosystems
It Came from Outer Space

This final chapter looks at the way in which narratives about vampiric invasions from outer space often work as a metaphor or mirror to illustrate either the self-protective qualities of the planetary ecosystem or as a more galactic idea of self-protection. They often tell of alien species coming to Earth because their own planet is exhausted or, conversely, focus on humans from Earth who make the reverse journey and thereby enact a form of ecological colonialism or biological energy vampirism. This group of films includes examples of more transient vampires that roam outer space looking for sustenance and acting as ecological, intergalactic regulators, as well as planets that lure the unwary into their clutches, both to feed themselves and to discourage the colonial intent of the Earth. Many of the narratives discussed in this chapter will naturally come from, or resonate with, the classic tale of space vampires, H.G. Wells' *War of the Worlds*, which will be the first text discussed below.

The War of the Worlds, H.G. Wells, 1897 / *Avatar*, James Cameron, 2009

Wells' novel tells of invaders from Mars who have expended the resources of their own world and who greedily eye the vigor and plenitude of Earth. Wells himself cites a conversation with his brother Frank over the English colonization of Tasmania off the coast of Australia, which resulted in the extinction of the indigenous population, as the original inspiration for the book. *The War of the Worlds* inveighs against the devastation wrought by colonial invasion on both the invaders and their victims, as well as the fear of reverse colonialism and invasion literature in general. Invasion literature was popular in the UK between 1871 and the outbreak of World War I, expressing the British fear of imperial decline where foreign powers—predominantly France and Germany—would penetrate and conquer the British Isles. By and large, the invasion narrative was a xenophobic genre that cemented popular notions of foreigners and outsiders as dangerous others. Wells' tale differs from both the earlier model of the invasion narrative and most of the alien invasion stories that followed his example, as it does not "culminate in a heroic confrontation between human beings and aliens in which the invaders are repulsed by the brave and ingenious actions of the human beings" (Alt 2014, 27). The Martians most decidedly fulfill this function of dangerous other, however, and *The War of the Worlds* further suggests that on top of familiar dangers, the unknown (from above) can intervene without warning or any chance of survival.

The story itself is so well known that it seems almost unnecessary to retell it here, but the most important point, at least for this reading, is that humanity is completely defeated by the invaders, who have far superior technology, weapons, and firepower, ultimately penetrating "humanity's impregnable conceit that it possesses the most advanced mind in the universe" (Renzi 2004, 112). Salvation only comes from Earth's ecosystem itself in the form of the common cold, against which the outsiders have no defense at all—this idea is also taken from imperial and colonial expansion, where indigenous populations often have no protection against the more common diseases of Western Europe. Within this framework, vampirism takes various forms but largely centers around the Martians.

The invaders from Mars are vampiric in many ways after exhausting the resources of their own world and preying on Earth. This is most obviously seen in their need of human blood to survive but also in their cyborg hybridity, as they need huge spacecraft (cylinders) and exoskeletons (tripods) to survive in the Earth's atmosphere and move in its gravity—it also is a critique of more advanced societies' technological prowess over less-developed cultures, as seen in much of Britain's colonial endeavors. The Martians are shown to be utterly dependent upon the machinery they use, not just as war vehicles with which to overcome the human resistance but as a means of separating their bodies from the alien ecosystem—this is possibly most dramatically seen in the rather bombastic film adaptation *Independence Day* (Emmerich 1996), where the aliens themselves are very small and completely encased in a large, tentacled exoskeleton. The narrator describes this evolutionary process quite clearly in *The War of the Worlds*, explaining how the Martians have "become practically mere brains, wearing different bodies according to their needs just as men wear suits of clothes and take a bicycle in a hurry or an umbrella in the wet" (Wells 1992, 81). This kind of evolutionary cyborgism, while conforming very closely to Stone's notion of a transformative vampirism, actually distances the Martians from their own ecosystem so that they have little connection to it other than in terms of their own subsistence. As the narrator further explains, the outsiders have "evolved" to a point where they cannot exist without their machines:

> Strange as it may seem to a human being, all the complex apparatus of digestion, which makes up the bulk of our bodies, did not exist in the Martians. They were heads—merely heads. Entrails they had none. They did not eat, much less digest. Instead, they took the fresh, living blood of other creatures, and injected it into their own veins [Wells 1992, 78].

This reveals the Martians as equally alien to their own world as to that of Earth, as they are no longer a part of their home world's ecosystem but merely a vampiric force that literally drains the blood out of the environment. Mars, it would appear, has been exhausted, and to paraphrase from the 1943 vampire film *Son of Dracula* (Siodmak 1943), they have come to a "younger ... stronger, more virile" land and they will "fasten on it and drain it dry, just as he [they] did his [their] homeland."

One may therefore argue that technology has turned the Martians into vampires, removing them from any meaningful connection to their own environment so that they only know how to destroy it to satiate their own needs and desires. They are a wantonly destructive force on Earth, as they wreak devastation everywhere they go in their hunt for humans as a food source.[118] Not only does their technological hybridity (vampirism) mark them as superior to humans and their technology but they also share a similar evolutionary arc where mastery over the natural environment is tantamount. However, the ending of the story clinches the point that a war with the environment is ultimately self-defeating. The ending further suggests humanity might follow a different trajectory from the Martians, as it is not solely dependent upon machinery or the ideology of superiority and total consumerism as are the Martians.

Curiously, the Martians actually bring some of their own environment with them: a red weed that thrives together with the Martians and dies together with them as well—oddly mirroring Dracula's necessity of having soil from his homeland with him. The vegetation, or red weed as it is called, is a "vivid blood-red tint" that alludes both to the color of Mars itself but also the necessity of blood for the Martians, oddly also making the plants vampiric in some way; the 1978 musical version by Jeff Wayne shows the weed as having claws and trying to consume one of the characters, and the 2005 Steven Spielberg film shows the plant as demarcating Martian-held territory and being fed on processed humans.

---

118. Roger Corman's *Not of This Earth* (1957) similarly shows an alien race coming to Earth to get human blood to save their world, though rather than launch an immediate invasion they send one agent as an advance party—they are humanoid in appearance and so are able to blend in more easily than Wells' Martians.

Chapter 5. Vampire Ecosystems 157

Tom Cruise as Ray Ferrier surveys a home away from home, free from ecological invaders, in *War of the Worlds*. Directed by Steven Spielberg (Paramount Pictures, 2005).

In *The War of the Worlds*, the red weed, which "grew with astonishing vigour and luxuriance" (Wells 1992, 80), succumbs "almost as quickly as it had spread. A cankering disease, due, it is believed, to the action of certain bacteria, presently seized upon it" (Wells 1992, 91). The plants' linkage to the invaders is confirmed, as it too lacked "a resisting power against bacterial diseases—they [terrestrial plants] never succumb without a severe struggle, but the red weed rotted like a thing already dead" (Wells 1992, 91). Thus, as the invaders, both flora and fauna, all die at the hands of the Earth's ecosystem, the narrator talks of humanity's place within it: "By the toll of a billion deaths man has bought his birthright of the earth, and it is his against all comers; it would still be his were the Martians ten times as mighty as they are. For neither do men live nor die in vain" (Wells 1992, 106), hoping that the separation between mankind and the environment can be repaired after this near-extinction warning.

Though mankind has survived through no good deed of its own, it has not learned the lesson of the story, namely that colonialism is not just about the violence wrought in other lands but about one's separation from one's own—Christina Alt notes that Wells' scientific romances of this period were largely pessimistic about "humanity, science and the future of the earth" (Alt 2014, 25). As the 20th century unfolded, technology was used not only to benefit man but also to damage large swaths of the planet. Furthermore, while technological developments are able to benefit the environment as a whole, technology is constantly directed at taking us further and further away from our lived experience of our own world.[119] This point is brought home in the next film, which reverses the positions of alien and self within Wells' narrative but retains the importance of technology as a means of aligning oneself with or alienating oneself from the environment.

In many ways, *Avatar* is a mirror image of Wells' novel, seeing humanity as the interplanetary aggressors who travel to alien worlds to "drain" them of their natural resources.[120] Though Mars is not the destination, forces from a dying Earth travel to a less technologically advanced planet called Pandora—it is actually a moon of a much larger planet called Polyphemus that is not unlike Jupiter in our own solar system—which is 4.3 lightyears away and is a rich source of "unobtainium," an element that provides huge amounts of power and sells for millions of dollars. In many ways, this intergalactic colonialism (Martians invading Earth) follows the example set by the British Empire (Britain invading Tasmania, among others), which Wells would have been familiar with, and more obviously that of America and global corporations in the 21st century that are driven by money rather than ecological or moral concerns—Bryn V. Young-Roberts suggests that it reflects the then–American government's operations in Iraq (Young-Roberts 2012, 18), and Francis Pheasant-Kelly emphasizes this, noting not only the reiterated use of the term "Shock and Awe" in the Iraq conflict (Pheasant-Kelly 2013, 161) and also in *Avatar* but also the film's director James Cameron's own words that make the connection between the invasions of Pandora and Iraq by the (American) military explicit (Hoyle 2009, in Pheasant-Kelly 2013). Not unlike Haddon's *Dracula*, there are echoes of the East India Company in that the military and mining operation on Pandora is run by a private group called the Resources Development Administration (RDA), and while they seem to have some form of authority bestowed upon them from Earth, they are very much focused on their own economic profit. The humans from Earth are thus framed as vampiric consumers, sucking out the planet's resources with

---

119. Not just in the continuing exploitation of the planet's natural resources and inability to responsibly cope with the waste produced but in the fascination with projects regarding space travel, such as Elon Musk's SpaceX project to literally separate humanity from the Earth and its expended resources.

120. Stories reversing the "Martians invading the Earth" framework of Wells' story appeared almost immediately, with Garrett P. Serviss' Edison's *Conquest of Mars* appearing in 1898 (although this was actually the sequel to Serviss' earlier book *Fighters from Mars*, which was an unauthorized and altered version of *The War of the Worlds*).

little to no regard for the indigenous peoples or the effects their operations will have on the environment, perceiving Pandora as a buffet of commodities to sell or buy.

The film shows the RDA as aggressors who forgo friendly, cooperative means of extracting the unobtainium and resort to military force to secure their assets. While the military campaign is waged, researchers hired by the RDA have cloned "avatars" of the Na'vi that can be controlled by human consciousness to better understand the inhabitants of the planet. However, Pandora's ecosystem fights back, as the indigenous people, the Na'vi, along with the avatars are joined by the planet's fauna in launching a victorious assault against the mining company and banishing them from their world. Not unlike in Wells' *War of the Worlds*, it is the connection between the Na'vi and their world that ultimately defeats the invaders, though as with the earlier story, technology plays a significant role in this—however, as with John Carter, there is much of the "white savior" motif in this final victory (see Birkin 2018, 31).

Although Pandora is a lush, verdant environment, its air is unbreathable to humans, forcing the invaders to rely on machinery and breathing apparatus to survive there. RDA's technology can be divided into two categories, which serve two purposes, respectively: one is to master the environment, and the other is to help the humans become more attuned to the ecosystem in order to try and forge a working relationship with the Na'vi for the purposes of safer mineral extraction. Each type of technology configures their respective users' approach to the environment of Pandora. As with the Martians in *The War of the Worlds*, the military and resource extractors use machinery to ease and facilitate their work, and even though the human operatives are shown at their physical peak, they merely act as "drivers" of the technological bodies they inhabit, effectively becoming vampiric cyborgs able to change their shape and their function as they move into different vehicles or robotic suits.

Two scenes exemplify this excessive reliance on technology, showing how the human invaders separate themselves not only from their environment but also from each other. The first is a meeting between the disabled veteran Jake Sully (Sam Worthington) and Colonel Quaritch (Stephen Lang) in the cargo bay of the main base on Pandora—Quaritch is an important character in the film, alongside Parker Selfridge (Giovanni Ribisi), who is the corporation's representative, as he embodies the brutality and objectivization inherent in any and all colonialist projects. Quaritch is meeting Jake as part of their agreement that Jake must keep the colonel informed of the progress of the avatar program in return for new mechanical legs. Quaritch himself sits in a huge cargo-robot, a massive mechanical skeleton that shifts heavy loads around the loading bay, which is both incredibly threatening in its physicality but also keeps its driver isolated and secure from the environment around them.[121] In contrast, just in front of this mountain of metal is Jake sitting in his wheelchair. The chair looks extremely basic for a culture that has so much advanced technology at its disposal and is nothing more than a chair suspended between two wheels, offering no sense of protection for its occupant and seemingly leaving him more vulnerable to his surroundings. The scene makes clear the otherness of Jake and the impossibility of him ever reintegrating into human society. Quaritch, for all his dependence on machinery, is also fixated on the integrity of the body, both in his own physical body—which he displays at every opportunity—and in his abhorrence of the avatar program, which separates the mental and physical selves and leaves the body an empty husk. Consequently, Jake will never be good enough for Quaritch, no matter what cyborg appendages might be applied to his body.

This same separation is seen in the second scene, which is nearer the end of the film, where Quaritch battles Jake's Na'vi avatar. Once again, the colonel is in a robot suit very similar to the one seen earlier, but rather than in the human, militarized environment of the RDA, it is in the Pandorian forest. After Jake's avatar has led the Na'vi to defeat the corporation's forces in the air, Quaritch seeks to enact revenge by turning off the life/transfer pod in the mobile laboratory where Jake's body resides while his mind is in the avatar. On his way to the laboratory, Quaritch is on the verge of killing Neytiri (Jake's Na'vi love interest), but Jake's avatar intervenes by attacking the human in the robot. The human's positioning is much as before, except that the environment around him is now hostile as well. The avatar, however, is totally at home, and although only armed with a pole, he manages to break the robot's plastic hood, allowing the atmosphere into the housing. Although Quaritch is still in charge of the metal skeleton of the machine, he is now vulnerable to the outside world in a way that the avatar is not. Just as the colonel is about to kill Jake's helpless body, the revived Neytiri shoots an arrow into his chest—arrows are always portrayed as "natural" weapons in the film and are used almost as part of a sacrament between the Na'vi and the animals/food they need to kill to survive[122]—revealing the sudden vulnerability of the exposed human body in this alien environment, particularly to the forces of natural protection created by the ecosystem itself.

In contrast to the military uses of technology, the researchers transfer their brain patterns into a genetically cloned body, or avatar, of a Na'vi—this

---

121. This speaks of the kinds of alienation that Karl Marx saw as inherent within capitalism. See Delaney and Reed 2013, 158.

122. It also suggests a rather problematic correlation between the Na'vi and Native Americans, and indeed African Americans, as both communities would appear to require the intervention of a "white savior" to gain agency, a point exacerbated by the casting of the film itself, where only African American and Native American actors were cast to play the Na'vi.

Chapter 5. Vampire Ecosystems

is a unique combination of human and Na'vi DNA—so that they can better understand, befriend, and become part of the local communities.[123] This transference oddly corresponds to the idea of a human consciousness acting as a "driver" within a technologically developed body that equally describes the military personnel, except of course that the robots inhabited by the military insulate them from the outside environment, whereas the avatar body inhabited by the researchers more strongly connects them to it. The role of the avatar body in connecting its host to the environment is particularly evident in the process undertaken by the disabled Jake. The disabled, "othered" human, who is rejected by the military (human) society, becomes fully other once in the avatar but is then accepted by the Na'vi. He befriends and later becomes the partner of Neytiri, who is a Na'vi princess. As he becomes more integrated into the community, Jake undergoes the tribe's rites of passage, which enable him to "bond" with a direhorse and a banshee—a process that entails the Na'vi to use their ponytails as "connectors" to similar appendages on the animals, forming some kind of empathic connection between them. Finally, Jake manages to mount and bond with a toruk—a huge and more vicious version of the banshee—a feat which is prophesied in Na'vi lore to unite the tribes under the leadership of the one who rides it.

Jake consequently becomes part of Pandora's ecosystem, no longer an outsider or foreign body but a part of its immune system. This point is brought home at the end of the film after Neytiri has killed Quaritch and saved Jake's suffocating human body. The Na'vi community decide to transfer Jake's consciousness to the avatar on a permanent basis using the power of the god Eywa—something of a correlation to Earth and the idea of Gaia—which utilizes the spiritual life-power of the ecosystem. This transfer marks a vampire-like transition, performing not only the kind of mind transference seen in films like *Daughters of Darkness* (Kümel 1971), *The Hunger* (Scott 1983), and *Nadja* (Almereyder 1994) but also that of the cyborg, bodily transformations earlier identified as inherently vampiric. Jake's human body dies, but his avatar body awakens to become part of Pandora itself.

The surviving humans, including the representative of the corporation, Selfridge—the king of the consumerist vampires, as it were—are forced to leave the planet, but the researchers in the avatar program are allowed to stay and become part of an environment where the ecological balance remains intact. The next films more clearly follow the Wellsian template of Martians and others from outer space as inherently dangerous and avaricious; indeed, the otherness of outer space itself is what makes them so, as here it is the alien worlds themselves that are dangerous.

---

123. Na'vi are blue, over 10 feet tall, have long tails, and are considerably stronger than humans.

It: The Terror from Beyond Space, Edward L. Cahn, 1958 / *Planet of the Vampires,* Mario Bava, 1965

It: The Terror from Beyond Space and *Planet of the Vampires* both follow expeditions outside of the confines of the Earth and, in the case of *Planet of the Vampires*, even the solar system. Neither of the movies represents its version of humanoid life as explicitly aggressive colonizers or exploiters of these undiscovered worlds, but rather the films caution against the dangers of the unknown and specifically of ecosystems that we do not understand. They thereby offer interesting correlations to the ideas of the blank whiteness/abyss of nature and the natural environment mentioned earlier as well the "darker" areas of the Earth's environment, such as the jungle from which many of the vampiric representations of the planet's eco-immune system are emitted. This intersection of the unknown (unrecognizable) and the dangerous other represents outer space as an inherently aggressive and unwelcoming environment, which for consumerist colonizers makes it even more deserving of exploitation and control. The hostility of the extraterrestrial environment to the human colonizers manifests itself particularly through organic or ecological agents. Not unlike *Avatar*, *It: The Terror from Beyond Space* and *Planet of the Vampires* utilize the extraterrestrial environment to enact something of a mirroring of the colonial policies of imperialism and capitalism on Earth by either bodily incorporating the other humanoids into its own ecosystem or, in an inversion of *The War of the Worlds*, killing the humans via processes of contagion and transformation.

It: The Terror from Beyond Space tells the story of the first manned mission to Mars in 1973 and its subsequent rescue mission. There is no explicit sense of a dying Earth nor any desperate need for resources, only a form of expansionism that is an accepted part of capitalism; while constructed as a benign enterprise, it is inevitably seen as a way to conquer or exploit other peoples/worlds/markets. The rescue ship arrives at the site of the first rocket, *Mars 6*, months after the rocket crashed, and apart from the debris, the landscape is largely desolate and rocky with little sign that anything is living or growing on it. However, the only remaining survivor from the rocket expedition, Colonel Edward Carruthers (Marshall Thompson), has a different tale to tell: "This was the planet Mars as my crew and I first saw it. Dangerous, treacherous, alive with something we came to know only as Death" (Cahn 1958). Carruthers has no idea what caused the death of his crew, only that something picked them off one by one and left behind no evidence. The rescuers find his story difficult to believe, particularly because they have discovered a skull with a bullet hole in it, but they nonetheless accept Carruthers

Chapter 5. Vampire Ecosystems 163

into their team and prepare for the journey home. However, unseen by the crew, a shadowy form enters the ship via an airlock, which is later sealed before taking off. Once they are back in space, curious things begin to happen on board, and when one of the crew goes to the lower decks to investigate, he is killed by a creature emerging from the shadows.

Carruthers thinks he hears something, and the rest of the crew soon join him in a search. Another of their number wanders off, and he too is killed by the thing from the darkness. Realizing they are now missing two members of the ship's crew, they initiate another search, and Carruthers discovers the first crew member's body in a vent. On pulling it out, they find that all his bones are broken and all the blood and fluid has been extracted from the body, leaving it shriveled and dry. Suspecting the other missing crewman might also be in the vent system, one of the men crawls inside and indeed finds him but is also attacked by the monster, though he manages to escape. Newly alert to the danger onboard, they seal up the vent and retreat to the upper deck of the craft. They try to variously blow up, shoot, electrocute, and irradiate the monster but to no avail. In one final attempt to destroy the creature and prevent its return with them to Earth, they don spacesuits and depressurize the spaceship through an airlock. This final resort seems to suffocate and kill the monster, making Earth safe once more.[124]

One of the most important aspects of the narrative is the creature's almost ghost-like quality that sees it disappear into and out of the shadows, which is overdetermined in the narrative by the actual size and physicality of the monster itself.[125] This relationship to shadows throws up some informative vampiric connections that intersect with the ecological facets of the story. Especially in the earlier parts of the narrative, the creature seems to exist entirely as an entity composed completely of darkness.

The monster thereby harks back to the earlier Count Orlok from Murnau's *Nosferatu*, who equally came from a mysterious, otherworldly land and was a creature of the darkness that could move across the landscape and enter any room as an almost supernatural shadow or, more accurately, an unholy absence of light. The monster from Mars is similarly elusive and insubstantial—when attacking the men of Carruthers' original expedition, it was as intangible as the mist that surrounded the Earthly invaders' vehicle. Stoker's *Dracula* takes on a similar form as a smote of dust. Furthermore, when the monster stows away on board the rescuers' ship, it does so only as a shadow. Onboard the ship, it moves through the air ducts and hides in the lower decks,

---

124. The ending does not make a huge amount of sense, as the creature obviously lived without trouble on the surface of Mars without oxygen.

125. As with many monster films of the 1950s and 1960s, the creature is a man wearing a large rubber suit, making its almost phantasmagorical nature in the story somewhat questionable.

Ray Corrigan as It appears as a monster of shadows in *It: The Terror from Beyond Space*. Directed by Edward L. Cahn (United Artists, 1958).

seemingly transmuting into the darkness—the ducts barely fit a human, let alone the hulking physical presence of the alien. The monster thus literally haunts the ship as the ghost of a dead world that it left behind.[126] The creature thereby becomes a manifestation of Mars itself, the specter of a world that has consumed itself and now lays in waiting for unsuspecting visitors. It is unclear whether there is more than one of the creatures, though potentially the planet might be able to bring them forth at will, but it embodies an insatiable hunger that will consume all that comes near it. The monster equally acts as a warning both of the results of insatiable consumption and of wantonly entering (colonizing) lands/environments/worlds that are not your own.

The monster is further able to transfer a specific form of infection, namely the contagion of insatiable consumption. The monster appears to kill all of its victims, enacting a kind of extreme vampirism in drinking/absorbing not only their blood but virtually all of their bodily fluids. However, one of the crewmen on the rescue ship is only wounded and manages to escape from the monster in the vents. Although the ship's doctor begins treating him, he quickly comes down with a fever that cannot be controlled. As the fever spreads, it becomes obvious that he will only survive with blood transfusions, curiously repeating the thirst of the monster itself. It is as though the creature has somehow infected the human or, more specifically, the human body with

---

126. Consequently, the film can be seen as a precursor to films such as *Alien* (Scott 1979), where the creature from another world haunts the humans' spacecraft as a ghost-like, chimeric presence that leaps out and takes the unsuspecting.

the same drive for blood that will eventually consume its own host,[127] again enacting the catastrophe that occurred on Mars. In a sense, the Martian is a form of traumatic supernatural contagion infecting all those that come into its orbit with the specter of irresistible, insatiable consumption that will reenact catastrophic ecological destruction over and over again. This reading casts new light on the ending, as the humans and the monster effectively destroy the very ship they need to survive, the monster in its destructive rage to reach the humans and the crew in their attempts to kill the monster before they reach the Earth. This struggle culminates in a scenario where the humans have no choice but to dispose of the one element that guarantees their own survival, oxygen, and to take it away long enough for the specter of Mars to die. The monster then is constructed as a vampire that survives on the energy of life itself, not just blood but all that provides the stuff of living matter; consequently, only the lifeless void of space can kill it.

The human crew eventually overcomes the vampire, leaving it to become part of the endless darkness of the void and ensuring that the Earth remains safe. *The Planet of the Vampires* is a similar tale and one where the home world is once again secure at story's end. Here though, the extraterrestrial vampire is not just an individual lifeform but the entirety of the planet itself.

The Planet of the Vampires is set in an unspecified future and the far reaches of outer space. Unlike in *The Terror from Beyond Space*, the spaceships are not on a purposeful expedition of colonization but are instead responding to a distress signal from an unknown planet.[128] The planet is "off the map," connecting this film to earlier travel narratives and adventure stories that not only discovered new worlds but also encountered myriad new dangers and cultures. Most of such journeys were funded by and founded on capitalist ideologies of exploitation and economic returns, with the "monsters" beyond the edge of the known conceived as existential threats to such endeavors. Indeed, this tenor of existential terror frames *Planet of the Vampires*, informing its depictions of acts of expansion into unknown environments and ecosystems as inherently colonial and carrying unimagined consequences. Bava's film resonates with later tales featuring apocalyptic forces that are released because of "civilized" travelers going into rainforests, jungles, or caves (as in the recent *The Silence* [Leonetti 2019]) in unexplored locations.

The effects of this uncharted planet begin to emerge before the expedition has landed—not unlike the environs of Orlok's/Dracula's castle, where the approach to the home of the vampire correlates with the journey becoming

---

127. The infection from the Martian very much mirrors that of the vampire, causing those affected to turn into vampires themselves, a point graphically shown in the film by a lingering shot on a large pile of empty blood transfusion bottles by the sick crew member.
128. This is a plot device that is now a staple of the genre and indeed of the film *Alien*, mentioned earlier.

increasingly more hallucinogenic—and the crews of both ships, the *Argos* and the *Galliott*, become hysterical and violent. Captain Markary (Barry Sullivan) of the *Argos* manages to regain control of his crew, but the Galliott crashes into the surface of the planet. Once the *Argos* lands, the crew goes to investigate the wreckage but finds everyone dead. Though they bury some of the crew of the *Galliott*, other bodies mysteriously vanish, and once the *Argos*' crew members return to their craft, peculiar things begin to happen. There are sightings of members of the *Galliott*'s crew alive and walking around outside of the *Argos*, and when the graves of the dead crew members are dug up, the bodies have gone, with only the sheets/shrouds they were buried in remaining.

The surface of the planet seems to be rocky and barren, not unlike that of Mars in *The Terror from Beyond Space*, with no apparent lifeforms present other than the colored lights flickering on and off in the swirling patches of mist covering the surface, yet the feeling that they are not alone persists, as is evident in this conversation between Markary and the ship's medical officer, Dr. Karan (Fernando Villena):

> MARKARY: One entire crew lost; two of our own crew gone. Bert dead, Eldon disappeared. And this unknown enemy keeps getting closer!
> DR. KARAN: The enemy is also becoming visible.
> MARKARY: What do you mean by that?
> DR. KARAN: Well, you saw something. Something not quite identifiable out of the corner of your eye.
> MARKARY: Ah, yes. As if it were composed of little globes of light, something fleeting, nothing definite. And the minute I looked at the things directly, they were gone [Bava 1965].

Markary's suspicions lead him to surmise that "if there 'are' any intelligent creatures on this planet ... they're our enemies," a view which is confirmed when they discover another ship on the planet, of which all the crew are calcified corpses.

Realizing they are not the first to have been lured into this trap, they rush back to the *Argos*, only to find more of their crew missing but also the members of the *Galliott* newly appearing, now no longer trying to hide. The dead bodies are in fact inhabited by alien entities whose world is on the brink of extinction due to its sun dying out, and they are consequently looking for a new home. However, they need bodies to inhabit so they can live in a different atmosphere—this world seems to allow them to exist as a miasmic entity within a host body—though when one of their number tears his space suit, the flesh below is shown to be rotten. The captain is horrified by this discovery and threatens that he and his crew will kill themselves to thwart the aliens' plan—not unlike the action taken by the crew of *Mars 6* in *The Terror from Beyond Space*. While this threat seems to work in the short term, it is not long before there is a fight between the aliens and the surviving astronauts, and three of the *Argos'* crew manage to escape into the ship. Two of the surviving crew members realize that Markary is behaving oddly and confront him, but not only is the captain an alien, so too is one of the other two. In the ensuing struggle, the last survivor disables the ship's defense shields before dying. Realizing they can no longer reach the home world of the *Argos'* crew members, the two aliens decide to reroute to the nearest planet, a rather backward one known as Earth. The twist thereby reveals that the humanoid space travelers were not humans at all but rather another alien race and that the space vampires are now on their way to consume our world.

This throws up some interesting constructions of ecosystems and the entities that can exist in them. In *The War of the Worlds* and *Avatar*, the invaders require a level of separation from the alien world they land on, but *Planet of the Vampires* complicates this. It is not clear whether they are a naturally vampiric, parasitic race that moves from host to host—not unlike the psychic parasites called Souls in *The Host* (Niccol 2013)—or if they were originally spirits/souls that were expelled from the dead bodies of aliens by the planet itself (the idea that land will expel those that are not welcome, not holy, etc.). However, their inherent vampiric nature of transmutation seems to allow them to exist as concentrated clouds of energy within the mists that swirl across the surface of the planet, as though the vampires are in synergy with the lifeless ecosystem—the calcified corpses in the other alien ship are only ever identified as victims and not necessarily the sources of the entities. That said, the vampires' connection to the dead world makes them representative of it, a manifestation of an ecosystem's undead drive to survive at any cost, even to the point of killing any entities that would deny it that right.

As seen in Bava's film, though, the vampires rarely remain satisfied on their home world and need to travel to others to continue feeding. The humanoid crew in *Planet of the Vampires* would appear to be unfazed about landing on the Earth, suggesting that the bodies of their new hosts will provide adequate organic exoskeletons to survive this new world, not that dissimilar to the Na'vi avatars from Cameron's film. While they perform a similar function to the robot machines of Wells' Martians, they do not so much protect them from the new ecosystem as integrate them into it. This would seem to be a similar outcome in Bava's film in that the aliens will be "cloaked" when entering the environment, getting past the ecosystem's natural immune system to then do as they please. This forms the basis of the next two alien invasions looked at here, with *Lifeforce* seeing aliens using human form to get to the Earth, while *The Thing* goes for something more fundamentally biological.

Lifeforce, Tobe Hooper, 1985 / *The Thing*, John Carpenter, 1982

In many respects, *Lifeforce* resembles *Planet of the Vampires*, likewise depicting an alien species that arrives on the Earth ready to drain humanity of its life-force. Although, as with *Planet of the Vampires*, we know very little about the aliens' home world, the aliens of *Lifeforce* likewise have the ability to take on any form they choose and inhabit the bodies of others. The original astronauts of *Lifeforce* are definitely human and investigating Halley's Comet, which passes in view of the Earth every 75 to 76 years. However, once in range of the comet, they spot a huge spacecraft hidden in its tail, and the crew proceed to investigate. Once inside the craft, they discover hundreds of large, desiccated, bat-like creatures floating in its main structure, but as they venture further, they find three naked humanoid figures—one woman and two men—floating in containers resembling glass coffins. The leader of the expedition, Colonel Tom Carlsen (Steve Railsback), becomes increasingly fascinated with the female figure. Indeed, all the male crew seem unable to take their eyes off her, describing her as "perfect."[129] The crew take the three caskets back to their ship but then suddenly lose contact with Earth.

When the ship returns to Earth's orbit and docks in space with another ship, the entire crew are discovered dead, except the three aliens who still seem to be in suspended animation, with evidence that there has been a fire onboard. The bodies are taken down to Earth to the Space Research Centre in London, and that is when things start to go dramatically wrong. The aliens are not dead, and the female revives, draining the life-force out of all the humans she encounters, leaving them as dried-up, undead husks. These husks, in turn, revive after two hours and need to suck the life-force out of another victim. Furthermore, to evade capture, the female alien transfers her consciousness into other bodies and hides her own body for future reclamation. The escape pod from the original spacecraft is concurrently discovered, containing an exhausted but still alive Colonel Carlsen, who is immediately brought to London. He then discovers that he has a special link to the alien woman, which makes him vital to the mission to track the vampire down and kill her.

When the female alien escapes from the research center, she overpowers her guards, not by force but by sexual attraction, as recounted by Dr. Bu-

---

129. This idea of sexual attraction to a female alien as a means to overpower humanity was also used in *Queen of Blood* (Harrington 1966). Here, the last surviving member of an alien race makes contact with the Earth and is picked up by a rescue ship on Mars. However, once on board the female alien begins to exert a strange sexual influence on the largely male crew and drinks their blood when alone with them. It transpires that she is actually a plant-based lifeform and that her own blood (chlorophyll) does not coagulate like that of humans, so when she is scratched by the only female member of the crew, she bleeds to death.

Chapter 5. Vampire Ecosystems 169

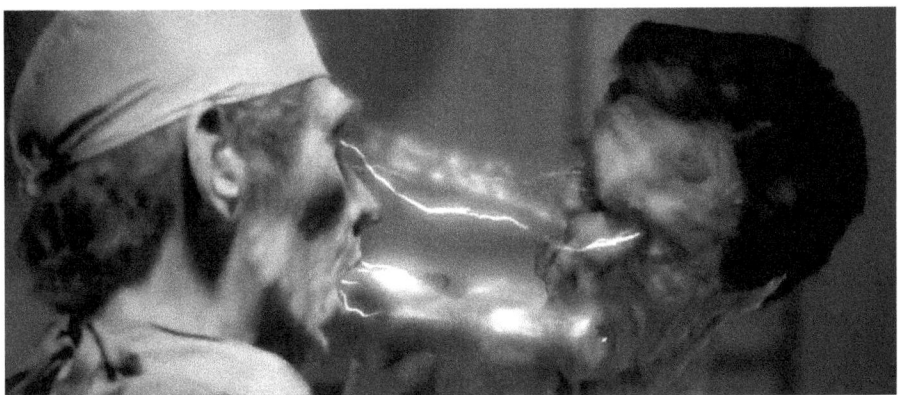

Ecological jouissance passing into the vampire with Jerome Willis as Pathologist (left) and Chris Jagger as First Vampire in *Lifeforce*. Directed by Tobe Hooper (Cannon Films, 1985).

kovsky (Michael Gothard), who survived the encounter, when questioned by the lead investigator Colonel Caine (Peter Firth):

> CAINE: Tell me again how the girl overpowered you.
> DR. BUKOVSKY: She ... was the most overwhelmingly feminine presence I have ever encountered. I was drawn to her on a level...
> CAINE: Was it sexual?
> DR. BUKOVSKY: Yes. Overwhelmingly so, and horrible. Loss of control.

This notion of alien sexuality, and female alien sexuality in particular, is not uncommon in later alien invasion films, such as *Species* (Donaldson 1995), *Species II* (Medak 1998), *Species III* (Turner 2004), *Species: The Awakening* (Lyon 2007), and *The Astronaut's Wife* (Ravich 1999). *Species* especially features a similar concept of an alien creature taking on the form of a beautiful and (often) naked human woman solely for the purposes of "glamouring" and procreating with a human male, with her true form—most clearly seen in *Species II* (Medak 1998) when mating with one of her own kind—being an excessive explosion of life and organic energy, not unlike that seen in *Annihilation*. In *Lifeforce*, however, the female alien is shown to have bonded with Carlsen even before he entered the vampire ship, reading his mind not only to learn more about his species but to bond with him personally and become his perfect mate (this resonates strongly with Ava from *Ex Machina*). Thus, once she is taken to Carlsen's ship, he is unable to resist her and consequently disables the ship's communications and kills the crew—he later blames the alien presence controlling him. He subsequently sets fire to the ship in a desperate bid to free himself from her and save his planet, as he tells one of the investigators, "Part of me didn't want to leave. She killed all my friends and I still didn't want to leave. Leaving her was the hardest thing I ever did." The female alien and the colonel share energies, and he carries part of her within him, creating a bond that continually draws him to her. He can likewise sense when she is near and can also see through her eyes—this actually corresponds to Stoker's *Dracula*, where the Count shares bodily fluids with Mina Harker, and afterward they are intimately and psychically connected to the point that she can see through his eyes. Carlsen uses this ability to help locate the alien woman, but she quickly moves through various human hosts and then travels as an energy beam back to her own body just in time for the alien ship to leave the comet's trail and begin orbiting the Earth directly above London.

Carlsen then realizes that he was being used by the female alien, and while she was leading him and Caine on a chase around England, the two male vampires had been infecting people in the capital, turning them into voracious, sexual/energy-seeking vampire/zombies. London consequently collapses into chaos and destruction—a scene that recalls the earlier film *Rabid* (Cronenberg 1977), where sexual vampirism and ecological imperative collide to unleash a biological jouissance (a rabid zombie horde) that threatens civilization itself.[130] The aliens actually use this process to drain the infected humans of their life-force, which is correlated within the film to the idea of a "soul," though this energy transference also equates to an extreme form of human interaction where, as explained by Dr. Fallada (Frank Finlay) of the research center, "in a sense we're all vampires. We drain energy from other life forms. The difference is one of degree." Fallada would seem to be referring to humanity's use of animals and the environment, with humanity as the pinnacle of such a vampiric hierarchy.

The aliens are thereby constructed as the "alpha" vampires, and when their victims die—they only ever attack humans—a ball of blue energy leaves the shriveled human body and either enters the alien or, as in the final stages of the film, is directed by the body of the female vampire up into the sky to the huge spacecraft that is hovering over London. At film's end the female vampire is killed—by leaded iron stake thrust through her power center just below the heart—halting the flow of energy to the alien ship, but hundreds of thousands of people have died, and the city is in flames. Once the energy stream is lost, the alien spacecraft leaves its orbit and powers into space, heading back toward Halley's Comet, suggesting that the aliens might return the next time the celestial body moves into view in 75 years' time,[131] which also

---

130. Stacey Abbott links the vampire infection in *Lifeforce* to the anxiety around the AIDS epidemic occurring around the time of the film's release (Abbott 2007, 138).

131. Comets have often been seen as portents of Earthly disasters, and horror films share in this symbolism of impending catastrophe, as seen in the novel and various screen adaptations of *The Day of the Triffids*.

implies that this might not have been the first time they have visited or collected energy from the inhabitants of the Earth.

In fact, Dr. Fallada from the center, who declares himself a thanatologist, suggests that folkloric tales of vampires might actually derive from earlier incursions of these aliens into our world. Consequently, the recurring appearances of the space vampires constructs them as some form of intergalactic regulating force that is allowed to "cull" the Earth's population at various intervals to keep their wanton expansionism in check and rebalance the planetary ecosystem. Something of this idea where space vampires are released at particular intervals to regulate the human population of the Earth frames the next film, though this alien is one that has been kept dormant under the ice caps until humans have developed sufficiently to be able to release it.

The Thing has been made into several films that follow much the same story but with important differences, and it is the development of these changes, especially between the 1951 version, *The Thing from Another World* by Christian Nyby, and the 1982 version, John Carpenter's *The Thing*, that is of most interest here. All the adaptations feature an alien spaceship crashing into the Arctic Circle and subsequently becoming buried in polar ice as the heat of the crash melts the snow, which then refreezes above it. An unsuspecting party of explorers/scientists discovers the wreckage and manages to retrieve the body of an alien—it is never suggested that more than one alien is on the ship—and take it back to their camp. Once there, the alien thaws rather more quickly than they anticipate and then begins to run amok in the camp, killing many of the occupants and ultimately threatening to overrun the entire population of the planet if allowed to escape.

In both the 1951 and 1982 adaptations, the alien craft appears to have crashed quite recently, but the fact that explosives are required to reach the ship beneath the ice suggests that it's buried in time, as though it is an object from the past, from the Earth itself, as much as from space. One may thus read the spacecraft as an object that should have been left alone, and the technological capability and expertise needed to unearth it are in themselves evidence enough that humanity has reached a point that requires an evolutionary regulatory intervention. In both the 1951 and 1982 versions, the uncovered alien can be read as a manifestation of the Earth's ecosystem, a released biological jouissance that is designed to aggressively regulate an imminently out-of-control humanity.

The 1951 film depicts the monster, not unlike the one in *The Terror from Beyond Space*, as a large, clumsy creature when visible and rather more nimble and anxiety-inducing when unseen. However, the monster's link to a bio-power beyond the merely human is discovered when one of the creature's arms is severed during its attack on the camp's sled dogs. On further investigation, the arm appears vastly dissimilar to human biology, lacking any nerves, and is actually plant-based and composed of chitins—which comprise the exoskeletons of crustaceans and beetles—making it extremely resilient. Furthermore, the severed limb seems to be revived or revitalized by blood, and after absorbing some blood from one of the sled dogs, it seems to act as a separate organism, able to exist autonomously without any connection to the body from which it was detached; in the meantime, the alien body is growing a new arm to replace it. The scientist examining the arm also notices some small seed pods attached to it, indicating that the creature can reproduce extremely quickly and in great number, which is concerning in many respects for the personnel of the remote outpost, not just because they will very quickly be outnumbered but because the creatures seem to prefer human blood as a food source—several of the scientists in the camp are found hanging from a ceiling beam with their throats cut to harvest their blood.

One of the scientists, in an attempt to understand the "visitor from another world," begins to grow some of the seed pods in an enclosure, feeding them small portions of human blood taken from the medical supply. Within four hours they already sprout and pulsate and, when using a stethoscope, can be heard making crying noises like hungry babies. Combined with the monster's ability to survive in extreme cold as well as in fire, this new species seems positioned to overtake and replace the human race on Earth, creating an entirely plant-based ecology—the creature has already consumed the pack of sled dogs on the base, so all fauna would appear to be a possible food source for it.[132] This is of course unacceptable to the human survivors on the base, apart from the one scientist who grew the pods, and so they conspire to find a way to kill it. As intimated earlier, they try to burn it but to no effect and damage parts of the camp while doing so, so they try electricity, which reduces the creature to ashes. The movie suggests that all evidence of the creature has thereby vanished, as the germinated pods in the laboratory were destroyed by fire earlier on and the alien spaceship seemed to evaporate when explosives were used to uncover it from the ice. This ending reinforces the idea of the creature as an ultimately insubstantial manifestation of the Earth's ecosystem testing itself against humanity, but it does seem too neat an ending, even for a Cold War–era film where monsters are more politically ideological than they are ecological. The 1982 film adaptation, *The Thing*, redresses this flaw in its predecessor, with the difference being that in this film the monster is not plant-based but rather a biological life-force capable of copying all types of organic life.

---

132. There is much here of the later *Invasion of the Body Snatchers* (Siegel 1956) in which alien pods land on the Earth and create perfect replicas of humans to replace them. This would also see the alien race as plant-based organisms, though this is never explicitly stated in the film.

Chapter 5. Vampire Ecosystems

Carpenter's *The Thing* begins differently from its precursor, not least as it is set in Antarctica rather than in the Arctic. We also see a spaceship entering the Earth's atmosphere at the opening of the film—we do not know if it lands or not, though it is later suggested that the alien spaceship found beneath the ice has been there for at least 100,000 years—and as the narrative starts the creature has already been released from its icy coffin and begun its rampant destruction. This is still unknown as the story opens, where we see a sled dog running toward and taking shelter in an American research base, Outpost 31—this is after the surviving two members of a nearby Norwegian expedition have failed to kill it. Later that night the American researchers put the new dog in the animal pen with their own pack. The dogs begin to react aggressively toward the newcomer, whose outer skin then begins to tear and transform, sprouting huge tentacles before attacking and consuming/absorbing the other animals. The American crew arrives just in time to see part of the creature detach itself and climb up into the ceiling and disappear, leaving the bulk below, which is then set on fire with flamethrowers. The camp's doctor studies the charred body of the creature, which is now a combination of several dogs entangled in a large, fleshy mass that defies explanation, exceeding the kinds of biological excess seen in *Annihilation*.

As the doctor peels back the handfuls of fleshy matter, the bones revealed beneath are a conglomeration of the animals it has assimilated pulled, stretched and twisted together in startling new configurations and with sets of long, sharp teeth where mouths once were. Marie Mulvey-Roberts cites its horrors as arising from its substance as "an undifferentiated body or interstitial creature," quoting Hurley's idea of the "gothicity of matter" (Hurley 1996, 33). Roberts further observes that the monster's "otherness, liminality and monstrosity" reaffirm the specificity of the human subject (Mulvey-Roberts 2016, 84). However, one might argue that the creature's grotesqueness shows the "stuff" of humanity to be essentially energized biological matter. This is equally horrifying and exuberant, so beyond human conceptions of nature that it is incomprehensible, gesturing toward life without restrictions and limits.

The camp's doctor concludes that the creature is able to transform itself into anything that it absorbs/eats, a theory he later confirms by studying the creature's cells under an electronic microscope that shows these cells consuming and then transforming into those of its new host/victim.[133] Using the same computer, the doctor predicts that the creature, which is correlated to a contagion, will have assimilated/infected the entire population in 27,000

---

133. Utilizing the now comparatively simple computer graphics of the times to display the cells attacking and consuming, they look not unlike the Atari arcade game Pac-Man.

An exuberance of biological life in *The Thing*. Directed by John Carpenter (Universal Pictures, 1982).

hours, or just over three years, if let loose into the world.[134] The inhabitants of the camp are further shocked by the realization that there is a 75-percent chance that one of them is already infected and is an alien creature pretending to be human. This fear is realized as they discover more than one of their number has been taken over by the creature, and the infected crew members variously have their chests torn open like huge mouths, biting off people's arms, or their heads grow spider-like legs and scuttle off across the floor.

As with the 1952 adaptation, the crew deduces that the Thing can exist not only as one entity but as many thousands, all able to contaminate any matter with which they come into contact, even down to the blood that comes from a torn-open body. The remaining survivors decide it is too dangerous to let the monster leave the camp, and they cannot let it freeze in the snow because then it could wait for rescuers to arrive and infect them, so they set fire to the base. The fire leaves two lone survivors sitting exhausted in the snow and staring at each other, each waiting to see if the other is infected.

The creature is a fascinating manifestation of unstoppable biological life and the planet's ecosystem ridding itself of a dangerous pestilence: humankind. It is definitely vampiric in the way it consumes/assimilates its victims—presaging other transformative monsters such as the T-X in *Terminator: Rise of the Machines* (Mostow 2003)—and in its ability to mutate and transform into any shape it chooses. In this sense, it is uniquely and unstoppably cyborg,

---

134. The first cases of what would later be called the AIDS "epidemic" were reported in June 1981, beginning the anxiety over contagion on a global scale, linked directly to air travel and the fact that the infected looked exactly like everyone else. Part of this was also the construction of the concept of "patient zero" and the idea of an originating point, whose discovery could control/cure the effects of the outbreak (see Wald 2008).

as it makes all appendages and additions part of its own body, existing both singularly and as a multitude. The creature is not explicitly plant-based, as in the earlier film, but it likewise possesses the ability to assimilate and copy any cell with which it comes into contact. Now that it has been unleashed, this excessive force is intended to return humankind to a level where they no longer threaten the Earth, or indeed the galaxy beyond that, and it is this last scenario to which the next two films look.

Solaris, Andrey Tarkowsky, 1971, and Steven Soderbergh, 2002 / *Event Horizon*, Paul Anderson, 1997

Event Horizon and both versions of *Solaris* deal with human emotion, in part because of its prominence within arguments around the uniqueness and superiority of humanity but also due to its ability to provide a form of psychic sustenance, at least in terms of certain approaches to vampire lore. In each of these films, the respective alien ecosystem specifically induces an affective condition in its human subjects/victims, though in two oppositional extremes, which might be called exhaustion and exuberance, in order to feed off of them and, more precisely, to prevent them from taking their consumerist excess out into the wider cosmos.

The first film, *Solaris*, has two versions that largely tell the same story of an alien world that holds the human explorers in orbit above its surface in a dream-like thrall to drain them of emotional energy; here, we will focus on the 2002 version but with recourse to the earlier one to fill in some detail that the latter narrative neglects. Soderbergh sets his tale in an undetermined future where space travel is commonplace but the Earth seems exhausted in an intermingling of the worlds of *Blade Runner* (Scott 1982) and *Wings of Desire* (Wenders 1987), mixing a constantly overcast, rainy future with the unbearable weight of memory and loss of post–World War II Berlin.[135]

The main protagonist, psychologist Chris Kelvin (George Clooney), who is recovering from the loss of his wife Rheya (Natascha McElhone), is invited by a friend, Gibarian (Ulrich Tukur), to come up to the space station *Prometheus*, which is orbiting the planet Solaris, to investigate recent unusual occurrences. We are never told how far the planet is from the Earth, but it does not seem to take Chris long to reach it, yet by the time he arrives, his friend has already committed suicide. The exploration of Solaris in this ad-

---

135. Both *Blade Runner* and *Wings of Desire* have vampiric aspects; in the first, the cyborg nature of the creation and existence of the replicants is performatively vampiric, while in the second the Angels that inhabit Berlin are actually energy vampires.

aptation is a purely commercial affair, not unlike that seen in *Avatar*. As Gordon, one of the few surviving members on the space station, tells Chris, she was sent there to "assess the economic potential of Solaris and whether or not it was a viable commercial property or energy resource." However, the planet has other ideas and reacts as a singular, connected environment that reaches out and reacts to lifeforms within its orbit. Tarkowsky's adaptation shows the planet surface covered in a vast ocean and swirling fog—not dissimilar to the mists and lights seen in *Planet of the Vampires*—which the crew of the space station describe as the planet's cerebral systems, parsing changes in its environment as the equivalent to thoughts. Soderbergh's film never shows the surface of the planet, only the opalescent atmosphere that changes in color and activity as the film progresses, beginning as blue with a few flashes of storms/thoughts across its surface and becoming purple to red and filled with lightning strikes as the craft, and Chris, descend into its midst.

In many respects, Solaris corresponds to the kind of thought transfer/psychic connection seen in *Lifeforce*, and at times the blue of the human life-force in *Lifeforce* looks very similar to that around Solaris, reinforcing the idea that it is life-energy of some kind. Not unlike Dracula in Stoker's novel, the planet is already reading Chris' mind to see how best to manipulate and control him, and this information finds solid form after his first night on the Prometheus. The planet connects more strongly to its human subjects once they are asleep, and as Chris dreams of his dead wife, he awakens to find her next to him in bed. Chris finds difficulty accepting her sudden appearance, even when the two surviving crew members of the *Prometheus*, Gordon and Snow (Jeremy Davis), explain the situation. Rheya is a newly created entity, probably constructed from subatomic particles that can think and feel like a human but are virtually indestructible—though they discover a way to make the aliens disassemble. Although she can think for herself, she is created from Chris' vision of her and has no memory of her life before waking up in his bed. Still distraught, Chris thinks he can dispose of the creature and ejects it into space, but when he awakes the following morning she has returned once more. Chris begins to accept her, but when Rheya discovers that he has already tried to kill her once before, she tries to commit suicide by drinking liquid oxygen.[136] In true vampiric fashion, she heals in front of his eyes and comes back to life.

The emotional tensions begin to escalate on the *Prometheus*, as reflected in the changing surface of the planet. When Chris first arrived, the few remaining crew members had reached a level of exhaustion and resigned acceptance of their situation, Gordon had disposed of her "visitor" by vapor-

---

136. It is suggested in the film that because Rheya committed suicide in real life and is the focus of Chris' memories of his wife, the new creation is predisposed to repeating this.

Chapter 5. Vampire Ecosystems 177

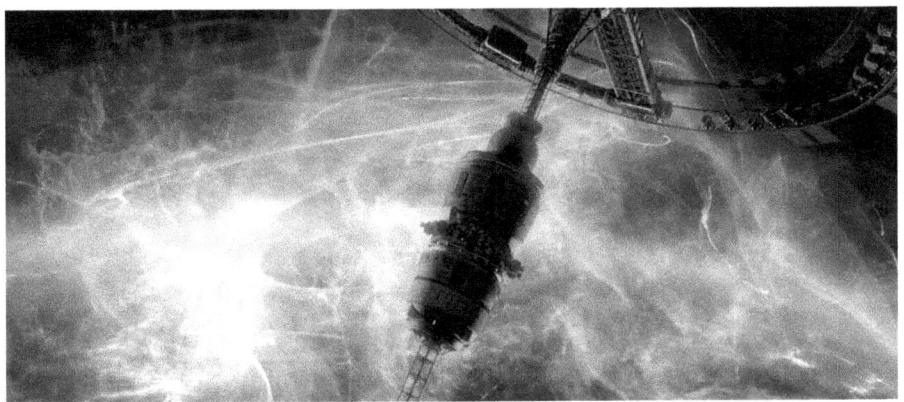

Solaris drawing emotional energy from the crew of the *Prometheus*. *Solaris*, directed by Steven Soderbergh (20th Century–Fox, 2002).

izing them—a process that deconstitutes the alien manifestations by using high-energy radiation—and Snow appears unaffected beyond his very vague manner of speech; both of them remain locked in their respective rooms. Their resignation kept the planet's surface at a calm blue, but when Rheya appears, Chris' emotional response begins to cause turbulence on the ship and the planet, and by the time Gordon has vaporized Rheya and they discover that Snow is actually a visitor, the planet has absorbed enough energy to expand and enclose the *Prometheus* in its atmosphere, which is glowing a hot purple/red. Not unlike Pandora in *Avatar*, the vampiric, consumerist humans are themselves the victims of a vampiric planet that is set on consuming them, though the exact nature of what this might be is not made clear.

As the ship's orbit disintegrates, both Gordon and Chris race for the escape pod to leave for Earth, but Chris stops on the verge of entering the pod as memories of Rheya seem to crowd his mind. Suddenly he appears to be back on Earth, as though the journey to Solaris had ended months before, and he is trying to reintegrate back into normal life. However, something is not quite right, and when he realizes he is reliving a scene that took place just before his flight to Solaris, he seems to awaken to find himself sitting on the floor of the *Prometheus* as the craft is falling into the planet. A little boy stands before him, a copy of Gibarian's son, who has been seen running around the ship throughout the film and who holds out his hand for Chris to take hold of. The moment Chris grasps it, he is back with Rheya, looking into her eyes and accepting any fate that means he can assuage his guilt over the past, as shown in this final exchange:

> CHRIS: Am I alive or dead?
> RHEYA: We don't have to think like that anymore ... we're together [Soderbergh 2002].

In a sense, Chris has been exhausted by his memories of the past, a never-ending nostalgia that provides sustenance for the planet—an idea reminiscent of *Wings of Desire*—providing something of an emotional, psychic power source for the alien world to live on. This idea is also seen in Tarkowsky's *Solaris*, in which Chris (Donatas Banionis) is about to leave the space station and we suddenly cut to an idyllic scene of him at his parents' cottage, which occurred before he left for space and where he is with his father once more—he always felt guilty about not spending enough time with his father. As the camera pulls back, it reveals the scene to be happening on an island on Solaris, where the astronaut will be with his father forever, slowly feeding the world that supports the dream.

Solaris replicates the Venus flytrap quality of Orlok in *Nosferatu* as an ecological honeytrap for human consciousness, molding itself to the observer's desires. Here, though, it can also be read that it is the exploitative intentions of the space travelers (the metaphorical flies) that provoke a certain kind of reaction from the planet. The Soderbergh film suggests more strongly than Tarkowsky's that the effects of Solaris might not be limited to the planet alone and, once "awoken," its vampiric tendrils could extend back to the colonists' home world. At one point, Chris suggests taking Rheya back to Earth with him, but Gordon is adamant this should not be allowed to happen, vehemently declaring, "She is not Human! Now I don't know about you, but I feel threatened by that" (Soderbergh 2002), as it is a mistake to think the planet's intentions are benign. Gordon further asserts that the humans have been manipulated by the alien world from the moment they arrived, suggesting that any emotionally influenced decisions they make are suspect.

Both film adaptations of Stanislaw Lem's *Solaris* establish very dreamlike qualities in their respective environments from the very outset. Each world is strangely soporific, full of loss and nostalgia, and the various flashbacks during the respective narratives seem to link it as a continuous vision that loops back on itself at the story's end. A line that appears in both films is "We don't want other worlds, we want mirrors," seeing our quest into outer space as a search to connect with ourselves; the vampiric planet fully understands this and induces a world that is like a never-ending dream that begins and ends with Solaris.

The next film works on a similar level, featuring an entity that utilizes the liminal states between the real and the imaginary to sustain itself, though here the world is not a dream but a never-ending nightmare.

Event Horizon is a far darker film than either of the *Solaris* adaptations, and it does not involve a planet but rather the greediest of celestial phenomena, a black hole, and still more, a manmade one. Unlike the *Solaris* films, *Event Horizon* has a very specific setting that is laid out at the start of the film in bold text across the screen:

Chapter 5. Vampire Ecosystems

> 2015
> First Permanent Colony Established on Moon.
> 2032
> Commercial Mining Begins on Mars.
> 2040
> Deep Space Research Vehicle "Event Horizon" Launched to Explore Boundaries of Solar System.
> She Disappeared Without Trace Beyond the Eighth Planet, Neptune.
> It is the Worst Space Disaster on Record [Anderson 1997].

As with *Solaris* (2002), *Avatar*, and the *Alien* series, among many others, space travel and exploration are predominantly seen as private-sector endeavors in the future, and while backed by governments (not unlike the East India Company mentioned earlier), they are essentially acts of colonization and exploitation. The *Event Horizon* has a very special place within this thematic, as behind the public mission mentioned above, it was an experimental vessel that could create its own wormhole on board that would allow it to travel vast distances in very short periods of time.[137] This would in turn open up the entire universe to commercial projects and potentially unlimited financial rewards. The vast commercial and monetary rewards to be gained by this venture are highlighted in the conversation between Dr. Weir (Sam Neil), who invented the gravity drive on the *Event Horizon*, and Lt. Starck (Joely Richardson), who questions the reasons for building such a device with little or no concern for the dangers involved:

> WEIR: When they [the three magnetic rings on the drive] align it creates an artificial black hole. Which allows the ship to travel to any point in space.
> STARCK: A black hole, the most destructive force in the universe. And you've created one?
> WEIR: Absolutely, Yes [Anderson 1997].

The creation of the spaceship *Event Horizon* is an extraordinary example of mankind's hubris, one that epitomizes a thoughtlessly aggressive thrust into worlds other than our own, so much so that it cannot go unchecked by the wider universe. As further noted by crewman Smith (Sean Pertwee) after an attempted attack on Weir, "When you break all the laws of physics, do you seriously think there won't be a price?"

The film follows the events that unfold after the missing *Event Horizon* suddenly reappears near Neptune in 2047, and a rescue ship, the *Lewis and Clark*, is sent to investigate. As soon as the rescue crew boards the formerly missing vessel, strange things begin to occur—not unlike those seen in

---

137. This is explained in the film as the wormhole being able to join two points lightyears apart in space by folding space itself so that the points are on top of each other, allowing the ship to pass through the hole almost instantaneously.

Solaris—and it is not long before members of the Lewis and Clark begin to die and the crew is forced to abandon their own ship to seek safety on the *Event Horizon*. The ship itself seems to be alive, reacting to the actions of the new arrivals—again not unlike Solaris responding to the increased emotional states of the crew of the space station. Things go from bad to worse, as the gravity drive seems to be the source of the ship's vitalization and a gateway to a vampiric dimension—the black hole itself—that feeds on the extreme emotions provoked in the crew by entrapping them in a never-ending nightmare.

In many respects, the *Event Horizon* is a futuristic nightmare of science, metal, and technology diametrically opposed to more natural and organic ways of life. However, the Gothic-like, steampunk vision of the gravity drive and the chamber around it shifts this perspective, framing the *Event Horizon* as a source of biological excess that cannot be resisted. The drive is made of three concentric metal rings around a large globe where all parts can rotate around a central core that sits in a large pool of water. The sphere is covered in dimpled disks that give it the appearance of a huge, metallic seed pod, and these shapes are mirrored in similar discs on the surface of the spherical chamber containing the drive. On the outer ring of the drive are spikes, not unlike the rear ends of beetles, that are repeated regularly on the walls of the chamber. The total effect is that of a cross between a thistle head and a beetle colony—a dark vision of irrepressible nature in a world of ice and metal.

Indeed, the drive responds to the crew members from the *Lewis and Clark* as if they were an alien presence, trapping them in various parts of the *Event Horizon*, causing individual members to have visions, and expelling the *Lewis and Clark* from its docking station. The black hole resembles oily-black fluid metal between the rings that pulses through the craft like a ripple or wave of water. It surges through the long central corridor of the *Event Horizon* as an organic representation of biological excess, causing severe damage to the rescue ship. Weir decides to investigate the cause of the drive's malfunction, but to do so he needs to access the computer panels in the walls of the chamber. As Weir opens the inspection hatch, bright green light spills out into the room. As he enters the access tunnels, all surfaces radiate the same luminous green color, and as the various shafts intersect and stretch out behind him, they resemble an arterial system filled with chlorophyll, providing life to the body of the ship and the drive itself, operating as a pulsing heart. The organic nature of the ship is remarked upon by the crew, as in the following exchange between Starck and the captain of the rescue team, Miller (Laurence Fishburne):

> STARCK: I think that there's a connection between the readings and the hallucinations [the crew have been experiencing] like, like they're all part of a defensive reaction, some sort of immune system.... I'm saying that this ship is

Chapter 5. Vampire Ecosystems

Jack Noseworthy as Justin in *Event Horizon* is confronted by organic machinery and a gateway to biological jouissance. Directed by Paul W. S. Anderson (Paramount Pictures, 1997).

> reacting to us and the reactions are getting stronger. It's as if the ship brought something back with it, a life force of some kind.
> MILLER: What are you telling me? That this ship is alive? [Anderson 1997].

This connection to the organic sees the ship act as both a focus and a manifestation of a wider ecological force seeking to limit the reach and interference of mankind. The *Event Horizon* is not unlike Solaris in its ability to sense and connect with humans in its vicinity, making them have dreams or hallucinate—Peter Wilberg talks of the *Event Horizon* as a never-ending dream within a dream leading to an ultimate awareness (in this case, the excess of the universe itself; Wilberg 2009). As with Solaris, the ship is able to read its victims' minds and consequently heighten their emotions, causing what Gavin Hurley calls "orgiastic gratuity" (Hurley 2017, 86), allowing it to feed off such psychic excess. It thereby facilitates the more explicitly vampiric wormhole that progressively glamours, lures, and cajoles the humans into becoming its longtime food source.

In fact, Weir begins to feel the effects of the *Event Horizon* while a long distance away from the ship, not unlike Chris in *Solaris* and Ellen in *Nosferatu*. Not long after leaving Earth, the crew of the *Lewis and Clark* are put into stasis in fluid tanks, and Weir is no exception, but he soon seems to enter a nightmare in which he encounters his dead wife—she committed suicide just as Chris' did—but her eyes are missing. As he struggles to escape her, he awakens in the tank and almost chokes, but it is opened just as they arrive at Neptune.

Indeed, Weir seems more affected than the rest of the crew, but it is not long before they all begin to suffer from waking nightmares of people they have lost or deaths they have caused. As events unfold, the nightmare becomes more and more real, and Weir is lost to the state of continual excess—the nearer they draw to the vampiric portal, the more ripples of jouissance are released, resonating more strongly with those who are connected to it. When Miller finds Weir alone in the control room, Weir has gouged out his own eyes and has dried blood over his face, and he tells the captain about "his" ship:

> WEIR: I created the *Event Horizon* to reach the stars, but she's gone much, much farther than that. She tore a hole in our universe, a gateway to another dimension. A dimension of pure chaos. Pure … evil. When she crossed over, she was just a ship. But when she came back … she was alive! Look at her, Miller. Isn't she beautiful?

As the film reaches its conclusion, the wormhole has opened and Weir invites Miller and the surviving crew to join him—by now, the ship's creator is largely naked and bald with his body covered in large cuts. Miller refuses but offers himself in return for the lives of the two other survivors, Starck and Cooper. Weir declines, knowing they all belong in the vampiric dimension, but unbeknownst to him explosives have been set to sever the command deck from the rest of the ship so it can act as a rescue pod of sorts. Some time later, the craft nears Earth and a rescue team boards it, releasing Starck and Cooper from the stasis pods. As Starck struggles to catch her breath, one of the rescuers removes their helmet, revealing Weir's mutilated face. Starck screams and re-awakens but this time in the arms of the real rescuers. Cooper approaches to calm her, but as he does so, the automatic doors close of their own accord, suggesting that the nightmare is not over. As with *Solaris*, the film is open-ended with no assurance of whether the nightmare is beginning or ending, nor does it clarify whether the influence of the *Event Horizon* is limited only to its close proximity—that is, Starck and Cooper never made it back to Earth—or has now come to the Earth itself to wreak havoc and despair, draining the world of human life in its exuberant excess of biological jouissance and restoring galactic balance once again.

The final two films discussed in this chapter both develop this notion of vampirizing the world of humanity while also restoring an ecosystem to balance.

Jupiter Ascending, The Wachowskis, 2015 / *The Day the Earth Stood Still,* Scott Derrickson, 2008

Jupiter Ascending and *The Day the Earth Stood Still* both represent Earth as a tiny cog in a much larger machine but one that has rusted to a point that the bigger forces at play in the wider universe need to repair it. In many ways, the higher powers in these visions of the cosmos are reflections of similar forces already at play on the Earth, though *The Day the Earth Stood Still* in particular moves beyond consumerist or economical considerations and views the balance of extraterrestrial powers purely through ecological eyes. *Jupiter Ascending* and *The Day the Earth Stood Still* are set in the present-time of filming but a version thereof that is highly futuristic in mise-en-scène, purposely showing Earth as one of the least developed planets, its humans as the most barbaric among the species that inhabit the solar system/universe.

Jupiter Ascending expands and contracts its lens to focus on both present-day America, specifically Chicago, and huge spaceships and planetary mining facilities well beyond the limits of Earth and draws tantalizing parallels between the two settings, suggesting that both are highly non-ecological environments. Chicago, representing 21st-century America, is shown as a glittering, shimmering landscape of lights towering into the night sky but one devoid of greenery or connection to the land and the countryside outside of it—very similar to the Los Angeles portrayed in the *Blade* series of films (*Blade* [Norrington 1998], *Blade II* [del Toro 2002], and *Blade: Trinity* [Goyer 2004]), as so too do the monumental space stations of the Abrasax family and the Aegis (space police) seem to be sparkling cityscapes floating in the blackness (blankness) of space. Both worlds seem focused on consumption and social hierarchies as Jupiter Jones, a young Russian émigré, oscillates between them, revealing their underlying resemblances.

The story focuses on a family of immortal, intergalactic royalty called the Abrasax that travel the universe fashioning new worlds so that they can "harvest" them at some point in the future to create an energizing elixir that keeps them young and provides huge economic rewards. The Earth is one of the harvested planets, a particularly valuable one that is owned by the matriarch of the family. The story opens with her murder, upon which she is subsequently reincarnated—a "recurrence," as it is called in the film—in the body of Jupiter Jones (Mila Kunis). In an attempt to get the rights to the Earth for themselves, the two Abrasax brothers, Titus (Douglas Booth) and Balem (Eddie Redmayne), try to, respectively, marry and kill their reincarnated mother. Jupiter manages to escape Titus and kills Balem in self-defense before returning to Earth and her old life but now as its ruler/owner. The Abrasax are explicitly called vampires in the film, a name seemingly given to them for their extremely long lives and consumption of humanoid life essence to remain young—though it is inferred that they are not the only race that might use this elixir—but the name equally applies due to their exploitation and manipulation of nature, constructing them, not unlike Wells' Martians, as metaphors for contemporary human disregard for their own ecosystem.

The House of Abrasax is one of the most powerful alien dynasties in the universe, which, since the death of the matriarch, is presided over by Balem, Titus, and Kalique (Tuppence Middleton). They are an extremely long-lived species—Kalique is 14 millennia old, and the matriarch was over 91 millennia old when she was murdered/killed. That said, they do age, and it is this part, rather than their great longevity, that more clearly establishes their vampiric credentials, as they replace dead cells in their body with those found in an elixir made of the life-force of humans. This is explained by Kalique to Jupiter as the immortal steps out of her rejuvenating bath looking like a 20-year-old again:

> Each of us has a code for our optimal physical condition. The problem is our genes have an expiration date which is transferred to our cells. A long time ago, someone figured out how to replace deteriorating cells with new ones. Today, it's as easy as changing a light bulb [the Wachowskis 2015].

Jupiter asks whether clones can be used for this process, but apparently they cannot, as Kalique explains, "No. Clones lack genetic plasticity. Several million years ago, a gene plague caused by cloning, nearly annihilated the entire human race." Consequently, only humans can be used to manufacture this elixir, and so the Abrasaxes have "seeded" many worlds across the universe that they can subsequently "harvest." Once a planet's population has increased to the point that the planet is unable to sustain it, it is considered ripe for harvest of this resource. As Kalique explains further to Jupiter, "Your earth is a very small part of a very large industry."

Indeed, harvesting is run as a huge and efficient factory process, as Kalique says to Titus in an earlier scene, "I've heard they feel no pain. It's all quite humane, from what I've been told," to which her brother responds, "Well ... there are Marshalls and administrators to make sure everything is done according to code ... but still ... it can be rather ... affecting."[138] In a later scene on Balem's processing planet, the so-called "affecting" dimension of this process is dramatized as a conveyor belt carrying a multitude of humans passes beneath the elder Abrasax brother—they are literally "beneath his heel," as the process occurs directly under a glass floor upon which Balem walks—and various whirring metal instruments extract pieces of skin and brain tissue to test for quality, only for the inert, paralyzed figure to drop through to a more invasive process. This resonates with other vampire texts in which humans are "farmed" to provide a steady supply for the vampiric overlords, as seen in *Daybreakers*, *Matrix*, and *The Strain*, though as mentioned it is not so much a food source in *Jupiter Ascending* as an elixir or power-smoothie to revitalize themselves.

---

138. This can of course be correlated to humanity's "harvesting" of animals for food, which is equally labeled as "humane" but often shown to be anything but. The film does not make the connection explicit, though it is worth noting that Lena Wachowski is a vegan.

Chapter 5. Introduction

The selection and development of "stock" is equally vampiric, as the Abrasax can only harvest planets that are genetically compatible with their own genes—if humans are seen as "natural" in this sense, then the vampires/Abrasax are supernatural in that they are human but also more than human. Their super-humanity is intimated by the fact that, as mentioned by Kalique, Jupiter could partake of the same process to lengthen her own life—but they happily splice, hybridize, and mutate other species to perform specific tasks for them. Caine (Channing Tatum), who saves Jupiter, is humanoid spliced with a wolf to be a better hunter/tracker for military use, and Stinger (Sean Bean), Caine's former captain, is crossed with a bee to heighten his sense of loyalty and teamwork—bees on Earth have also been genetically modified to recognize Abrasax royalty.

The universe beyond Earth thus becomes a vampiric environment of transformed and transforming beings that can become whatever the Abrasax desire, yet Earthlings, or normal humans, are kept purposely ignorant of the universe beyond their planet, even when said universe bursts into their everyday lives. Thus, in Jupiter's conversation with Kalique, when she asks the immortal if she is part of a vampire race, Kalique replies, "We are the cause of a lot of those myths, but my mother was as human as you or I." This secrecy is further brought home in the aerial battle scenes as Caine saves Jupiter from the various factions of hunters trying to kill or capture her on behalf of Balem or Kalique. As the pursuit dives and swoops around the skyscrapers of nighttime Chicago, the buildings are fired upon, exploded, set on fire and generally destroyed. Narrowly escaping, Caine and Jupiter leave the city by car early the following morning, and she asks her protector how the Abrasax will explain all the damage and still keep alien life secret. Caine tells her to look in the rear-view mirror, and already the fires are dying out and buildings are being rebuilt. He explains that by the evening all will have returned to its previous state, and by then agents of the vampires will have fixed the memories of human witnesses to the events. Caine points out to Jupiter that already she cannot explain the picture of a Keeper on her phone and that she perceives the events of yesterday as a dream—a particularly impressive feat in the age of smartphones and ubiquitous CCTV.[139] This resonates with vampire texts such as *Dark City*, in which a race of alien parasites changes the cityscape and the inhabitants' memories every night to persuade them that all is well, as well as more Earthbound series like *Moonlight* (Koslow 2007–8), *Being Human* (Whithouse 2008–13) and *The Originals* (Plec 2013–18), to name but a few, all of which feature a hidden world of vampires that is kept secret from the wider population by using glamouring and/or cleanup teams.

---

139. There is the inevitable nod to people claiming alien abduction and the like here when Caine explains that those the agents cannot find are considered mad.

The reconstructing of Chicago suggests that the wider Abrasax empire has a vested interest in keeping Chicago intact, reinforcing Chicago's role as a synecdoche of this empire. The process by which it is achieved further implies that the same kind of mechanisms—excessive consumerism and coercion—hold sway in Chicago and the wider empire. In the early stages of *Jupiter Ascending*, the eponymous Jupiter works for, and is exploited by, her Uncle Vassily—Vassily is in charge of the extended family's finances and employment opportunities but shows preferential treatment to his own children. This exploitation creates a hierarchy within her "human family" where those at the top metaphorically (economically) feed off those at the bottom. Indeed, her job as a maid cleaning rich people's apartments and toilets reinforces this point. Furthermore, early in the film her cousin Vladdie suggests that Jupiter sell some of her own eggs to a fertility clinic for money, introducing from the outset the notion of bodies as commodities. Interestingly though, money is not the most important aspect of this transaction, as Jupiter later learns from Kalique:

> KALIQUE: In your world, people are used to fighting for resources ... like oil, or minerals, or land. But when you have access to the vastness of space, you realize there's only one resource worth fighting over ... even killing for: More time. Time is the single most precious commodity in the universe [the Wachowskis 2015].

Jupiter's human and alien families alike consume humans in inferior positions and sell "time" to make money. It is not so surprising then that at the film's end, Jupiter does return to the vampiric world of Earth, a place that will happily consume itself before any alien race will have the opportunity to do so.

The last film discussed in this chapter does not replicate the excess of the Earth in space but rather sees the wider universe act to prevent it, and the vampire in it is accordingly very much the enforcer of such a galactic ecological decision.

The Day the Earth Stood Still is an adaptation of the 1951 science fiction film of the same name by Robert Wise and, like *The Thing*, is not necessarily an explicitly vampire text, yet there is much about the 2008 version that tends toward the vampiric. The original novel was very much about post–World War II America and featured an alien named Klaatu and his robot protector, Gort, who land in Washington, D.C., to warn representatives of the world that if they do not change their violent ways, which are beginning to pose a threat to their galactic neighbors, then the Earth will be destroyed. By the new millennium, mutually assured destruction was less in the popular consciousness than environmental disaster, and so in the 2008 adaptation the purpose of Klaatu's visit has changed slightly and is no longer about the damage humans might do to other planets but the damage that they are wreaking on their

Chapter 5. Introduction

own—Klaatu explains that the rarity of worlds such as the Earth is sufficient reason to destroy any threats to its continued survival. Consequently, the alien visit becomes not so much about changing the minds of world leaders but rather about Klaatu himself, who becomes an almost biblical figure in saving several ark-loads of species from the Earth before halting the final plague of pestilence to give humanity one last chance in the form of celestial brinksmanship. This theme of humanity changing for the better when it is left with no other choice was mooted in the 1951 version, which allows "humans one final opportunity," while the 2008 one "barely averts a preordained plan to annihilate the entire planet" (Pollard 2011, 146).

In the original version of the film, Klaatu (Michael Rennie) is simply a humanoid alien hailing from a considerably more technologically/morally advanced planet. In the 2008 version, he is far more exotic and vampiric in his ability to change his appearance, as well as in his ability to become part of and control the environment. The 1951 original opens with a brief scene set in the snow-covered mountains of north India, where a lone mountaineer encounters a large, glowing orb in the snow. He chips a hole in it with his pick and then loses consciousness, awakening to find the orb gone. The 2008 version features a much larger version of the globe which lands in Central Park, out of which an amorphous figure emerges, only to be shot by an over-excitable soldier. On further study in a secret government facility, researchers discover that the shape is largely made up of a substance resembling whale fat, which operates similarly to a placenta, and which then gives birth to the alien contained within it. The hairless, naked form that emerges is obviously copied from the mountaineer seen at the beginning of the film. As explained by Dr. Benson (Jennifer Connelly), this alien was literally born into our ecosystem in a form with the correct physiology to enable it to survive. Klaatu (Keanu Reeves) swiftly matures into an adult human, though it is suggested that he could take on the appearance of any lifeform and that he originally existed as a very different entity in his home world, as he tells Dr. Benson:

> KLAATU: This body will take some getting used to.... It feels unreal to me. Alien. It will take time to adapt.
> BENSON: What were you before you were human?
> KLAATU: Different.
> BENSON: Different how?
> KLAATU: It would only frighten you [Derrickson 2008].

His alien alterity further manifests itself in his ability to repair himself by rubbing a gel into his skin which repairs large wounds on his chest. He also seems to have the capability of revitalizing the recently deceased, and he can draw the pestilence from the bodies of Dr. Benson and her son into his own body. Klaatu thus emerges as both Christ-like and vampiric in his abilities—two figures that are not mutually exclusive in popular culture. His ability to transform his appearance and to control others—he directs operatives in the government facility using his mind—and to bring others back from the dead are all highly vampiric accomplishments.

Klaatu further exhibits the usual identifiers of the vampire as an outsider, or "other" from the land beyond the forest, and one who has much more affinity for the natural world than for the civilized peoples he encounters. The main issue in *The Day the Earth Stood Still*—as the vampire knows whereas the humans do not—is that the humans are an inveterately colonizing species obsessed with ownership. This point is raised quite early in the story when the representative of the American president, Regina Jackson, demands to know of Klaatu why he has come to *our* planet, to which he replies:

> KLAATU: *Your* planet.
> REGINA JACKSON: Yes; this is our planet.
> KLAATU: No, it is not [Derrickson 2008, emphasis in original].

Something of this attitude is seen in the preponderance of security and military personal and facilities that feature in the film (Dalby 2010, 114), revealing how much humanity (America) rigorously controls, regulates and exploits the lands, assets and resources it calls its own. Klaatu of course sees the Earth as part of a much larger ecosystem, a universal one where unruly and destructive races/species have no place, especially ones that refuse to mend their ways, as he explains to Professor Barnhardt (John Cleese), who is arguing for a second chance for humanity: "Your problem is not technology. The problem is you. You lack the will to change…. I cannot change your nature. You treat the world as you treat each other." Klaatu thereby forms a bridge between the Earth and regulatory forces of the universe, as part of both, whereas Gort, the robot that arrived with him, is the real vampire presence of the film.

In the 1951 original film, Gort is a rather static figure, a man made of large silver tubes who serves as an accessory to both Klaatu and the spacecraft. He has effective laser eyes, and although Klaatu suggests he is very dangerous, he seems more homely than homicidal. In the 2008 adaptation, the robot is nameless—the scientists investigating him call him GORT: Genetically-Organized Robotic Technology—and far more menacing. He is far larger than his earlier counterpart and sleekly matte black. Klaatu manages to prevent him from taking extreme defensive measures after he is shot while departing from his globe, but this Gort is more independent than in the original film. In the 1951 version it functions as the robotic wingman to a more humanoid (White) leader, but here it configures the main purpose of the visit from outer space. Once Klaatu is taken away, Gort then stands guard by the globe, but it is not long before the robot is shooting down missiles and aircraft that are trying to destroy it and/or the extraterrestrial ship. The military responds by constructing a kind of heavy-duty metal box around the robot, which curiously does not seem to trigger its defense mechanism, though it is just as likely that Gort is letting itself be taken to the secret compound because its removal suits its purposes.

Once in the facility, Gort begins to reveal its more vampiric characteristics as researchers fail to ascertain its robotic nature, as seen in the exchange between government agent Driscoll and the colonel in charge of the testing chamber:

> DRISCOLL: So, is it a machine or a living thing?
> COLONEL: Both. Or neither.
> DRISCOLL: What do you mean "neither?"
> COLONEL: It seems to be some sort of silicone-based hybrid. We're calling it GORT: Genetically Organized Robotic Technology [Derrickson 2008].

This positions Gort in very vampiric territory, being neither alive nor dead, a cyborg, a robot, and organic. Indeed, it seems to have the same kind of affinity to insects, bugs, and pestilence as Orlok and Dracula had. Gort thus transforms into an all-consuming cloud of insect-like nanorobots that can consume anything and everything, including humans, and as they do so they absorb that matter and use it to create more copies of themselves.

The resultant swarm of nanorobots/insects is as deadly as any plague and flows across the landscape, destroying all manmade artifacts. It is an all-consuming pestilence of biblical proportions, as suggested in the earlier scenes of alien globes across the world, collecting specimens of all the species of fauna from the Earth. This collection resonates with the story of Noah and the ark[140]—which intimates that Gort, as the Master Vampire, is the angel of death who follows upon the construction of Noah's Ark.[141]

The 2008 film also elects for a different setting from the original, and although they are both based in America, the 2008 version lands the spaceship, now a huge, shining globe, in New York, establishing a connection between Klaatu's quest and the movement of capital and consumerism in the Big Apple. Indeed, *The Day the Earth Stood Still* creates a world like that of *Jupiter Ascending*, where consumerism is inseparable from life in the 21st century. This idea explicitly runs throughout the story, from the blatant product placement—McDonalds in particular—and more interestingly as a sign of the imminent destruction of humanity. Consequently, toward the end of the film, when the "final plague" is unleashed, the first target is a huge hauling

---

140. This idea also features in the apocalyptic film *Knowing* (Proyas 2009) that came out the following year, but here, examples of humans are also taken, as the Earth itself will be destroyed by huge solar flares.
141. The series *The Strain* also uses the idea of the Angel of Death for its Master Vampire, though humanity just barely survives.

Vampiric Gort transmuting into a cloud of life-consuming nanorobots in *The Day the Earth Stood Still*. Directed by Scott Derrickson (20th Century–Fox, 2008).

truck that is literally consumed by the plague as it drives through an enormous power-generating plant. The entire scene is full of factories burning materials that produce black smoke from multiple chimney stacks, with electricity pylons everywhere, so much so that at one point it looks more like an oil field in the Middle East than a powerplant near New York City. This choice of a hauling truck as the first target of destruction does not seem coincidental, with the power industry producing much toxic waste and contributing to climate change, as does the transportation of food stocks in gas-guzzling lorries. The plague then proceeds to the New York Giants football field, another symbol of modern consumerism, marketing, and revenue production, which crumbles under the onslaught of destruction, destroying the entertainment industry, as was also seen in *The Omega Man*.

The cloud of pestilence finally reaches New York City, the center of consumerism and finance, where it is finally halted by Klaatu. This is the scene after which the film is titled, as Klaatu's destruction of the plague drains power/electricity from every network, appliance, and battery (in the 1951 original it was made clear this would not affect hospitals and similarly vital sites). In a series of cut shots, the story establishes the blackout as a worldwide phenomenon, after which it shifts from images of global capital cities to images of a large harbor where huge cargo ships are bestilled in the water, unable to dock. The scene then shifts to a car factory production line that has ground to a halt and shows workers walking away from the inert machinery—oddly reminiscent of *Christine* and its use of car production to symbolize human consumption. This final action suggests the kind of change that is required of humanity to save the Earth and themselves but equally implies that mankind would simply stop without the impetus of consumerism. This hiatus is left unresolved, as it is at that point that Klaatu returns to his globe-shaped

Chapter 5. Introduction

spaceship and departs the Earth, leaving the planet's inhabitants to seize the opportunities opened up by their new, de-capitalized reality or suffer the consequences.[142]

Ultimately, *The Day the Earth Stood Still* gives humanity too easy a pass. Nothing in the film has intimated that mankind will do anything differently in relation to the wider ecosystem, and what Klaatu took for signs of hope were acts of kindness between humans, not toward other species nor toward the planet as a whole. However, before the rather contrived ending to the story, Klaatu speaks more straightforwardly for the Earth itself as a planet that exists within the wider conditions of the planetary ecosystem but is not beholden to the cultural baggage and concerns of human history leading up to the 21st century. This sense of viewing the Anthropocene era from the outside gives the newcomer an element of clarity in regard to the importance of the relationship to one's environment and a life lived in balance and intelligence. As Klaatu explains to Dr. Benson:

> KLAATU: This planet is dying. The human race is killing it.
> HELEN BENSON: So, you've come here to help us.
> KLAATU: No, *I* didn't.
> HELEN BENSON: You said you came to save us.
> KLAATU: I said I came to save the Earth.
> HELEN BENSON: You came to save the Earth ... from us. You came to save the Earth *from* us [Derrickson 2008, emphasis in original].

This sums up the underlying theme of most of the films discussed in this study, where the most dangerous species on the planet is mankind, and the ecosystem of the forest, the landscape, the undeveloped world, the planet, the solar system, or the cosmos will do anything it can to protect itself from us. The most important part of this, in relation to this volume, is the place of literal and metaphorical vampires within these frameworks of resistance, renewal, and, sometimes, revenge. The vampire is thus not only a representative of the all-consuming abyss of nature but a vital means of considering the relationship between humanity and the planet. The figure of the vampire, in all its manifestations, symbolizes how out of balance we are with our environment. Since the 17th century and the outbreak of the vampire panics across Western Europe, the undead have emerged when things have gone wrong, when cultures clash, and when ideas of balance, synergy, and understanding are broken. More often than not, this centers on the clash between the old world and the new. On the larger historical scale, the vampire represents a past that is more connected to its roots and the environment, even if by superstition and folklore, as opposed to so-called more enlightened times of movement and change. On a more contemporary

---

142. Something not too dissimilar is seen in *The World's End* (Wright 2013).

and smaller temporal scale, the vampire points out the ways in which humanity is unwilling to change its selfish, consumerist "traditions" in favor of discovering new ways of living that conserve and share resources.

While a rather homogenous reading, and one that is beyond just the differences between the rural and the urban—though ongoing anxieties between these two spaces provide much vampiric energy to some of the texts mentioned here—it does capture something of many of the films discussed here. This study posits that these films fashion their vampires as "natural" responses to (modern) human intrusion, showcasing the disparity between developed and underdeveloped countries and the implications of colonialism and exploitation. Some of the more futuristic texts discussed herein often seem to endorse such echoes of the distant past, suggesting that the onset of modernism, capitalism, and consumerist ideology and its culmination in neoliberalism have caused humanity to lose an important essence in its existence. This is not necessarily a call for Marxist eco-warriors or environmental socialism, but it does posit that ecology and economics are not a zero-sum game and that the world will not stop turning if our focus returns to considering our place in relation to the Earth as our evolutionary home and partner. Many of the films discussed herein suggest that one does not need to know, or possess, all parts of one's home and that the homely (heimlich) and the unhomely (unheimlich) are not oppositional binaries or mutually exclusive. There is much to be learned from gazing into the abyss of nature, which we need not and cannot own, dominate, or possess. Its difference from us, as expressed through the body of the vampire (or vampiric body), is both functional and necessary, a symbol of the need for respect and a reminder of our own place within the ecosystem. Wells' story of the planet saving itself should be remembered, bearing in mind that the survival of humanity was not because of a special union between ourselves and the world we call home but a lucky coincidence. In defending itself, the world had not thought of our protection but was, like Gort, showing us what happens to those who do not belong and want to selfishly drain the Earth of all its resources. Indeed, all the various insects, bats, fungi, animals, spectral essences and vampiric excesses that have plagued mankind in the pages of this volume show us that we are not essential to the continuation of the Earth. Neither by divine right nor by evolutionary right, the Earth is not ours to do with as we please. We must earn our place within its ecosystem. Vampires serve as a timely reminder that to keep both the world and ourselves alive into the future, we do not need to accept the vampires but to recognize them for the existential threat they are (even the sparkly ones) and accept their challenge to be more at home in the world we live in.

Filmography

The Addams Family, dir. David Levy (Filmways, 1964–6).
The Addams Family, dir. Barry Sonnenfeld (Paramount Pictures, 1991).
Aliens, dir. James Cameron (20th Century–Fox, 1986).
Annihilation, dir. Alex Garland (Paramount Pictures, 2018).
The Astronaut's Wife, dir. Rand Ravich (New Line Cinema, 1999).
Aswang, dir. Wrye Martin and Barry Poltermann (Prism Entertainment Corporation, 1994).
Aswang, dir. Jerrold Tarog (Regal Entertainment Inc., 2011).
Aswang, dir. Michael Laurin (Distribber, 2018).
Autómata, dir. Gabe Ibáñez (Contracorrientes Films, 2014).
Avatar, dir. James Cameron (20th Century–Fox, 2009).
Back to the Future, dir. Robert Zemeckis (Universal Pictures, 1985).
The Bat People, dir. Jerry Jameson (American International Pictures, 1974).
Bats, dir. Louis Morneau (Destination Films, 1999).
Bats: Human Harvest, dir. Jamie Dixon (Sci-Fi Channel, 2007).
Being Human, created by Toby Whithouse (Touchpaper Television, 2008–13).
Billy the Kid Versus Dracula, dir. William Beaudine (Embassy Pictures Corp., 1966).
Bird Box, dir. Suzanne Bier (Netflix, 2018).
The Black Water Vampire, dir. Evan Tramel (Image Entertainment, 2014).
Blacula, dir. William Crain (American International Pictures, 1972).
Blade, dir. Stephen Norrington (New Line Cinema, 1998).
Blade II, dir. Guillermo del Toro (New Line Cinema, 2002).
Blade Runner, dir. Ridley Scott (Warner Bros. Pictures, 1982).
Blade: Trinity, dir. David S. Goyer (New Line Cinema, 2004).
The Blair Witch Project, dir. Daniel Myrick and Eduardo Sanchez (Artisan Entertainment, 1999).
Blood Car, dir. Alex Orr (TLA Releasing, 2007).
Blutgletschern [Blood Glacier, a.k.a., The Station], dir. Marvin Kren (Allegro Film, 2013).
The Book of Eli, dir. the Hughes brothers (Summit Entertainment, 2010).
Bordello of Blood, dir. Gilbert Adler (Universal Pictures, 1996).
Burnt Offerings, dir. Dan Curtis (United Artists, 1976).
Cabin Fever, dir. Eli Roth (Lionsgate Films, 2002).
Captain Planet and Planeteers, created by Ted Turner and Barbara Pyle (Turner Programme Services, 1990–96).

The Car, dir. Elliot Silverstein (Universal Pictures, 1977).
Chosen Survivors, dir. Sutton Roley (Columbia Pictures, 1974).
Christine, dir. John Carpenter (Columbia Pictures, 1983).
Curse of the Undead, dir. Edward Dein (Universal Pictures, 1959).
Dark City, dir. Alex Proyas (New Line Cinema, 1998).
Daughters of Darkness, dir. Harry Kümel (Showking Films, 1971).
The Day of the Triffids, dir. Steve Sekely (Allied Artists, 1963).
The Day of the Triffids, dir. Nick Copus (BBC, 2009).
The Day the Earth Stood Still, [a.k.a. Farewell to the Master, and Journey to the World], dir. Robert Wise (20th Century-Fox, 1951).
The Day the Earth Stood Still, dir. Scott Derrickson (20th Century-Fox, 2008).
Daybreakers, dir. the Spierig brothers (Lionsgate, 2009).
The Dead Undead, dir. Matthew R. Anderson and Edward Conna (Phase4Films, 2010).
The Devil Bat, dir. Jean Yarborough (Producers Releasing Corporation, 1940).
Devil Bat's Daughter, dir. Frank Wisbar (Producers Releasing Corporation, 1946).
Dinoshark, dir. Kevin O'Neill (Anchor Bay, 2010).
Dracula, created by Cole Haddon (NBC Universal Television Distribution, 2013-14).
Dracula III: Legacy, dir. Patrick Lussier (Dimension Films, 2005).
Dracula Untold, dir. Gary Shore (Universal Pictures, 2014).
Dracula's Daughter, dir. Lambert Hillyer (Universal Pictures, 1936).
El pueblo fantasma [Ghost Town], dir. Alfredo B. Crevenna (Studios América, 1965).
Event Horizon, dir. Paul Anderson (Paramount Pictures, 1997).
Ex Machina, dir. Alex Garland (Universal Pictures, 2014).
The Fearless Vampire Killers, or Pardon Me, Your Teeth Are in My Neck, dir. Roman Polanski (Metro-Goldwyn-Mayer, 1967).
Food of the Gods, dir. Burt I. Gordon (American International Pictures, 1976).
The Forsaken: Desert Vampires, dir. J.S. Cordone (Screen Gems, 2001).
The Frankenstein Chronicles, created by Benjamin Ross and Barry Langford (Rainmark Film, 2015-17).
From Dusk Till Dawn, dir. Robert Rodriguez (Miramax Films, 1998).
From Dusk Till Dawn, created by Robert Rodriguez (Miramax Television, 2014-16).
From Dusk Till Dawn 2: Texas Blood Money, dir. Scott Spiegel (Dimension Home Video, 1999).
From Dusk Till Dawn: The Hangman's Daughter, dir. P.J. Pesce (Dimension Home Video, 1999).
Frostbiten, dir. Anders Banke (Paramount Pictures, 2006).
Game of Thrones, created by David Benioff and D.B. Weiss (Warner Bros. Television Distribution, 2011-19).
Ganja & Hess, dir. Bill Gunn (Kelly Jordan Enterprises, 1973).
The Girl with All the Gifts, dir. Colm McCarthy (Warner Bros. Pictures, 2016).
Grizzly, dir. William Girdler and David Sheldon (Film Ventures International, 1976).
The Hallow, dir. Corin Hardy (Entertainment One, 2015).
The Happening, dir. M. Night Shayamalan (20th Century-Fox, 2008).
The Haunting of Hill House, created by Mike Flanagan (Netflix, 2018-present).
The Host, dir. Andrew Niccol (Open Road Films, 2013).

The Hunger, dir. Tony Scott (Metro-Goldwyn-Mayer, 1983).
The Hunger Games, dir. Gary Ross (Lionsgate Films, 2012).
Hybrid [a.k.a. Super Hybrid], dir. Eric Valette (Anchor Bay Entertainment, 2010).
I Am Legend, dir. Francis Lawrence (Warner Bros. Entertainment, 2007).
I Bought a Vampire Motorcycle, dir. Mont Campbell (Hobo Film Enterprises Ltd., 1990).
I Walked with a Zombie, dir. Jacques Tournier (RKO Radio Pictures, 1943).
Independence Day, dir. Roland Emmerich (20th Century–Fox, 1996).
Invasion of the Body Snatchers, dir. Don Siegel (Allied Artists Pictures, 1956).
Island of the Burning Damned [Night of the Big Heat], dir. Terence Fisher (Planet Film Productions, 1967).
Island of the Doomed [La isla de la muerte], dir. Ernst von Theumer (Orbita Films, 1967).
It: The Terror from Beyond Space, dir. Edward L. Cahn (United Artists, 1958).
Jaws, dir. Steven Spielberg (Universal Pictures, 1975).
John Carter, dir. Andrew Stanton (Walt Disney Studios Motion Pictures, 2012).
Jug Face [a.k.a. The Pit], dir. Chad Crawford Kinkle (Moderncine, 2013).
Jupiter Ascending, dir. the Wachowskis (Warner Bros. Pictures, 2015).
Jurassic Park, dir. Steven Spielberg (Universal Pictures, 1993).
Kingdom, dir. Kim Seong Hun (Netflix, 2019–present).
Knowing, dir. Alex Proyas (Summit Entertainment, 2009).
The Lair of the White Worm, dir. Ken Russel (Vestron Pictures, 1988).
Land of the Dead, dir. George A. Romero (Universal Pictures, 2005).
The Last Man on Earth, dir. Sidney Salkow and Ubaldo B. Ragona (American International Pictures, 1964).
Let the Right One In, dir. Tomas Alfredson (Sandrew Metronome, 2008).
Lifeforce, dir. Tobe Hooper (TriStar Films, 1985).
The Little Shop of Horrors, dir. Roger Corman (American International Pictures, 1960).
The Little Shop of Horrors, dir. Frank Oz (Warner Bros. Pictures, 1986).
Locusts, dir. David Jackson (Sci-Fi Channel, 2005).
Mad Max: Fury Road, dir. George Miller (Warner Bros. Pictures, 2015).
The Matrix, dir. the Wachowski Brothers (Warner Bros., 1999).
The Mist, created by Cristian Torpe (Spike, 2017).
Monstrum, dir. Heo Jung-Ho (Kidari Ent, 2018).
Moonlight, created by Ron Koslow and Trevor Munson (Warner Bros. Television, 2007–8).
Nadja, dir. Michael Almereyder (October Films, 1994).
Near Dark, dir. Kathryn Bigelow (DeLaurentiis Entertainment Group, 1987).
Night of the Living Dead, dir. George A. Romero (Continental Distributing, 1968).
Nightwing, dir. Arthur Hiller (Columbia Pictures, 1979).
Nosferatu, dir. F.W. Murnau (Fine Arts Guild, 1922).
Nosferatu the Vampyre, dir. Werner Herzog (20th Century–Fox, 1979).
Not of This Earth, dir. Roger Corman (Allied Artists, 1957).
The Omega Man, dir. Boris Sagal (Warner Bros. Pictures, 1971).
Orca, dir. Michael Anderson (Paramount Pictures, 1977).
The Originals, created by Julie Plec (Warner Bros. Television, 2013–18).

The Passage, created by Liz Heldens (20th Television, 2019–present).
Penny Dreadful, created by John Logan (Showtime Networks, 2014–16).
Piranaconda, dir. Jim Wynorski (Sunfilm Entertainment, 2012).
Planet of the Vampires, dir. Mario Bava (American International Pictures, 1965).
Primal, dir. Josh Reed (Primal Films, 2010).
Prowl, dir. Patrik Syverson (After Dark Films, 2010).
Queen of Blood, dir. Curtis Harrington (American International Films, 1966).
A Quiet Place, dir. John Krasinski (Paramount Pictures, 2018).
Rabid, dir. David Cronenberg (New World Pictures, 1977).
Rampant, dir. Kim Sun-hoon (Next Entertainment World, 2018).
The Red Violin, dir. François Girard (Odeon Films, 1998).
Revolt of the Zombies, dir. Victor Halperin (Academy Pictures Distribution Corporation, 1936).
Road, The, dir. John Hillcoat (Dimension Films, 2009).
Rosemary's Baby, dir. Roman Polanski (Paramount Pictures, 1968).
The Ruins, dir. Carter Smith (Paramount Pictures, 2008).
Salem's Lot, dir. Tobe Hooper (Warner Bros. Television, 1979).
Scream Blacula Scream, dir. Bob Kelljan (American International Pictures, 1973).
Shadow of the Vampire, dir. E. Elias Merhige (Lionsgate Films, 2000).
Sharktopus, dir. Declan O'Brien (Anchor Bay, 2010).
The Shining, dir. Stanley Kubrick (Warner Bros. Pictures, 1980).
The Silence, dir. John R. Leonetti (Netflix, 2019).
Snow White and the Huntsman, dir. Rupert Sanders (Universal Pictures, 2012).
Snowpiercer, dir. Jon-ho Bong (The Weinstein Company, 2013).
Solaris, dir. Andrey Tarkowsky (Mosfilm, 1971).
Solaris, dir. Steven Soderbergh (20th Century–Fox, 2002).
Son of Dracula, dir. Robert Siodmak (Universal Pictures, 1943).
Soylent Green, dir. Richard Fliescher (Metro-Goldwyn-Mayer, 1973).
Species, dir. Roger Donaldson (Metro-Goldwyn-Mayer, 1995).
Species II, dir. Peter Medak (Metro-Goldwyn-Mayer, 1998).
Species III, dir. Brad Turner (Metro-Goldwyn-Mayer Home Entertainment, 2004).
Species: The Awakening, dir. Nick Lyon (Metro-Goldwyn-Mayer Home Entertainment, 2007).
Splinter, dir. Toby Wilkins (Magnolia Pictures, 2008).
Stake Land, dir. Jim Mickle (Dark Sky Films, 2010).
Stake Land II: The Stakelander, dir. Dan Berk and Robert Olsen (Glass Eye Pix, 2016).
Stake Land—Mister, dir. Glenn McQuaid, *Glasseyepix*, 11 May 2011, https://m.youtube.com/watch?v=B12PrMFSg68&list=PL34A2B05A7EDD0627&index=6, accessed 21 August 2019.
Stake Land—Origins, dir. Larry Fessenden, *Glasseyepix*, 11 May 2011, https://m.youtube.com/watch?v=9EXSQRJ6XS4&list=PL34A2B05A7EDD0627&index=7, accessed 21 August 2019.
Stephen King's The Mist, dir. Frank Darabont (Dimension Films, 2007).
The Strain, created by Guillermo del Toro and Chuck Hogan (20th Television, 2014–17).
Stranger Things, created by the Duffer brothers (Netflix, 2016–present).
Surviving Evil, dir. Terence Daw (Kaleidoscope, 2009).

Teletubbies, created by Anne Wood and Andrew Davenport (BBC Worldwide, 1997–present).
Terminator 3: Rise of the Machines, dir. Jonathan Mostow (Warner Bros. Pictures, 2003).
They Have Changed Their Face, dir. Corrado Farina (Garligiano, 1971).
The Thing, dir. John Carpenter (Universal Pictures, 1982).
The Thing, dir. Matthijs van Heijningen Jr. (Universal Pictures, 2011).
The Thing from Another World, dir. Christian Nyby (RKO Radio Pictures, 1951).
Thirst, dir. Rod Hardy (New Line Cinema, 1979).
30 Days of Night, dir. David Slade (Sony Pictures, 2007).
30 Days of Night: Blood Trails (seven-part miniseries), dir. Victor Garcia (FEARnet, 2007).
30 Days of Night: Dark Days, dir. Ben Ketai (Sony Pictures Home Entertainment, 2010).
30 Days of Night: Dust to Dust (six-part miniseries), dir. Ben Ketai (FEARnet.com, 2008).
28 Days Later, dir. Danny Boyle (Fox Searchlight Pictures, 2002).
28 Weeks Later, dir. Juan Carlos Fresnadillo (Fox Atomic, 2007).
The Twilight Saga: Eclipse, dir. David Slade (Summit Entertainment, 2010).
Upir z Feratu [Ferat Vampire], dir. Juraj Herz (Cinefear Ostalgica, 1982).
Valkoinen Pura [The White Reindeer], dir. Erik Blomberg (Adami Filmi, 1952).
Vampire Bats, dir. Eric Bross (Sony Pictures Television, 2005).
Vampires, dir. John Carpenter (Columbia Pictures, 1998).
Voodoo Island, dir. Reginald LeBorg (United Artists, 1957).
War of the Worlds, dir. Steven Spielberg (Paramount Pictures, 2005).
White Zombie, dir. Victor Halperin (United Artists, 1932).
Wings of Desire, dir. Wim Wenders (Argos Films, 1987).
The Witch, dir. Robert Eggers (A24, 2015).
World War Z, dir. Marc Forster (Paramount Pictures, 2013).
The World's End, dir. Edgar Wright (Universal Pictures, 2013).

Bibliography

Abbott, Stacey. 2007. *Celluloid Vampires: Life After Death in the Modern World*. Austin: University of Texas Press.
Alt, Christina. 2014. "Extinction, Extermination, and the Ecological Optimism of H. G. Wells," in Gerry Canavan and Kim Stanley Robinson (eds.), *Green Planets: Ecology and Science Fiction*. Middletown: Wesleyan University Press, 25–39.
Anon. 4 June 2014. "The Witches Curse: Clues and Evidence," *Secrets of the Dead*. https://www.pbs.org/wnet/secrets/witches-curse-clues-evidence/1501/. Accessed 21 August 2019.
Arata, Stephen. 1996. *Fictions of Loss in the Victorian Fin de Siecle*. Cambridge: Cambridge University Press.
Ashlin, Scott. "Nightwing," *1000 Misspenthours*. http://1000misspenthours.com/reviews/reviewsn-z/nightwing.htm. Accessed 21 August 2019.
Auerbach, Nina. 1996. *Our Vampire, Ourselves*. Chicago: University of Chicago.
Bacon, Simon. 2018. *Gothic: A Reader*. Bern: Peter Lang.
_____. 2 August 2018a. "The Girls Have All the Gifts," *Peter Lang*. https://medium.com/peter-lang/the-girls-have-all-the-gifts-83f259bf505a. Accessed 21 August 2019.
_____. 2019. *Dracula as Absolute Other: The Troubling and Distracting Specter of Stoker's Vampire on Screen*. Jefferson: McFarland.
Bane, Theresa. 2010. *Encyclopedia of Vampire Mythology*. Jefferson, NC: McFarland.
Banita, Georgiana. 2015 . "The Spirits of Globalisation: Masochistic Ecologies in Fabrice Du Welz's *Vinyan*," in Anil Narine (ed.), *Eco-Trauma Cinema*. New York: Routledge, 146–63.
Bérard, Cyprien. 2012. *Lord Ruthwen ou les Vampires*, a.k.a. The Vampire Lord Ruthwen [1820], trans. by Brian Stableford. Tarzana: Blackcoat Press.
Birkin, Laura. 2018 . "*Avatar* (2009)," in Salvador Jimenez Murguía (ed.), *The Encyclopedia of Racism in American Films*. Lanham: Rowman & Littlefield, 30–1.
Bishop, Kyle William and Angela Tenga (eds.). 2017. *The Written Dead: Essays on the Literary Zombie*. Jefferson, NC: McFarland.
Blackwood, Algernon. 27 February 2011. "The Transfer" [1911], *Algernonblackwood.org*. http://algernonblackwood.org/Z-files/Transfer.pdf. Accessed 21 August 2019.
Boylan, Andrew M. 2012. *The Media Vampire*. Morrisville: Lulu.com.
_____. 27 November 2013. "Will the real Count Dracula please stand up?" *Taliesin Meets the Vampire*. http://taliesinttlg.blogspot.com/2013/11/will-real-count-dracula-please-stand-up.html. Accessed 21 August 2019.
_____. 5 April 2014. "Stoker and the Bat," *Vamped*. https://vamped.org/2014/04/05/stoker-bat/. Accessed 1 August 2019.
Browning, John Edgar. 2015. "Oil and the (Geo)Politics of Blood: Towards an Eco-Gothic Critique of *Nightwing*," in Katarina Gregersdotter, Johan Höglund, and Nicklas Hållén

(eds.), *Animal Horror Cinema: Genre, History and Criticism*. London: Palgrave McMillan, 94–109.
Burke, Edmund. 2015. *A Philosophical Enquiry into the Origin of our Ideas of the Sublime and the Beautiful* [1757]. Oxford: Oxford University Press.
Butcher, Daisy (ed.). 2019. *Evil Roots: Killer Tales of the Botanical Gothic*. London: British Library Publishing.
Byron, Glennis, and Aspasia Stephanou. 2013. "New-imperialism and the Apocalyptic Vampire Narrative: Justin Cronin's The Passage," in Johan Höglund and Tabish Khair (eds.), *Transnational and Postcolonial Vampires: Dark Blood*. London: Palgrave Macmillan, 189–201.
Christie, Deborah G. 2015. "Flickering Nitrate: The Cinematic Vampire as Social Other in *Nosferatu*," in Nadine Farghaly (ed.), *Beyond the Night: Creatures of Life, Death and the In-Between*. Newcastle-upon-Tyne: Cambridge Scholars, 265–82.
Clarke, Bruce. 2017. "Planetary immunity: Biopolitics, Gaia theory, the holobiont and the systems counterculture," in Erich Hörl (ed.), *General Ecology: The New Ecological Paradigm*. London: Bloomsbury Academic, 193–216.
Cooper, Ian. 2018. *The Manson Family on Film and Television*. Jefferson, NC: McFarland.
Copjec, Joan. 1994. *Read My Desire: Lucan Against the Historicists*. Cambridge: MIT Press.
Cothren, Claire Renae. 2015. *The Evolving Southern Gothic: Traditions of Racial, Gender, and Sexual Horror in the Imagined American South*. College Station: Texas A & M University.
Craft, Christopher. 1994. *Another Kind of Love: Male Homosexual Desire in English Discourse 1850–1920*. Berkeley: University of California.
Cronin, Justin. 2010. *The Passage*. London: Orion.
Csala-Gáti, Katalin and János I. Tóth. 2003. "The socio-biological and human-ecological notions in *The Time Machine*," *Wellsian: The Journal of the H. G. Wells Society*, 26, 12–23.
Curti, Roberto. 2017. *Italian Gothic Horror Films, 1970–1979*. Jefferson, NC: McFarland.
Dalby, Simon. 2010. "Apocalyptic Exceptionalism: Rosenberg, Clancy and the Prophecy of Americanism," in Jason Dittmer and Tristan Sturm (eds.), *Mapping the End Times: American Evangelical Geopolitics and Apocalyptic Visions*. Burlington: Routledge, 99–118.
Darwin, Charles. 1839. *Narrative of the surveying voyages of His Majesty's Ships Adventure and Beagle between the years 1826 and 1836, describing their examination of the southern shores of South America, and the Beagle's circumnavigation of the globe. Journal and remarks. 1832–1836*. London: Henry Colburn.
Days, Rhonda R. 2017. "Monsters of God: Negotiating the Sacred in Stake Land," in Cynthia J. Miller and A. Bowdoin Van Riper (eds.), *Divine Horror: Essays on the Cinematic Battle Between the Sacred and the Diabolical*. Jefferson, NC: McFarland.
Delaney, Tim and Ellen Reed. 2013. "A Marxist Look At *Avatar*," in Mary K. Leigh and Kevin K. Durand (eds.), *Marxism and the Movies: Critical Essays on Class Struggle in the Cinema*. Jefferson, NC: McFarland, 145–63.
del Toro, Guillermo and Chuck Hogan. 2009. *The Strain*. London: Harper.
_____. 2010. *The Fall*. London: Harper.
_____. 2011. *The Night Eternal*. London: Harper.
Demos, John Putnam. 2004. *Entertaining Satan: Witchcraft and the Culture of Early New England*. Oxford: Oxford University Press.
Eighteen-Bisang, Robert and Elizabeth Miller (eds.). 2008. *Bram Stoker's Notes for Dracula*. Jefferson, NC: McFarland.
Frayling, Christopher. 1992. *Vampyres: Lord Byron to Count Dracula*. London: Faber & Faber.
Gibson, Matthew. 2006. *Dracula and the Eastern Question: British and French Vampire Narratives of the Nineteenth-Century Near East*. London: Palgrave Macmillan.

Bibliography 201

Giesen, Rolf. 2019. *The Nosferatu Story: The Seminal Horror Film, Its Predecessors and Its Enduring Legacy*. Jefferson, NC: McFarland.

Glišić, Milovan. 2015. *After Ninety Years: The Story of Serbian Vampire Sava Savanovic* [1880], trans. by James Lyon. Charleston: CreateSpace.

Halberstam, Judith. 1993. "Technologies of Monstrosity: Bram Stoker's Dracula," *Victorian Studies*, 36/ 3 Victorian Sexualities, 333–52.

Hall, David D. 2004. *Witch-Hunting in Seventeenth-Century New England: A Documentary History 1638–1693*. Durham: Duke University Press.

Hantke, Steffen. 2019. "Alex Garland's *Ex Machina* (2014)—Science Fiction and Horror," in Simon Bacon (ed.), *Horror: A Companion*. Bern: Peter Lang, 111–18.

Harraway, Donna. 1990. *Simians, Cyborgs, and Women: The Reinvention of Nature*. New York: Routledge.

Höglund, Johan. 2014. *The American Imperial Gothic: Popular Culture, Empire, Violence*. New York: Routledge.

Hollm, Jan. 1999. "The Time Machine and the Ecotopian Tradition," *The Wellsian: The Journal of the H.G. Wells Society*, 22, 47–54.

Holloway, April. 3 September 2014. "Elizabeth Bathory—16th century deranged serial killer or victim of betrayal?" *Ancient Origins*.https://www.ancient-origins.net/history-famous-people/elizabeth-bathory-16th-century-deranged-serial-killer-or-victim-betrayal. Accessed 21 August 2019.

Horgan, John. 4 August 2015. "Bethe, Teller, Trinity and the End of Earth," *Scientific American*. https://blogs.scientificamerican.com/cross-check/bethe-teller-trinity-and-the-end-of-earth/. Accessed 21 August 2019.

Hoyle, Ben. 11 December 2009. "War on Terror backdrop to James Cameron's Avatar," *The Australian*. https://www.theaustralian.com.au/arts/film/war-on-terror-backdrop-to-james-camerons-avatar/news-story/9b5d2c1e310122f7370aac135194419e. Accessed 21 August 2019.

Hughey, Matthew. 2014. *The White Savior Film: Content, Critics, and Consumption*. Philadelphia: Temple University Press.

Hurley, Gavin F. 2017. "Nonknowledge and Inner Experience: A Post-Modern Rhetoric of Space Horror," in Michele Brittany (ed.), *Horror in Space: Critical Essays on a Film Subgenre*. Jefferson: McFarland, 81–95.

Hutchings, Peter. 2014. *The Horror Film*. New York: Routledge.

Jackson, Kevin. 2017. *Nosferatu (1922): eine Symphonie des Grauens*. London: BFI Film Classics.

Kawin, Bruce F. 2012. *Horror and the Horror Film*. London: Anthem Press.

Keetley, Dawn and Angela Tenga (eds.). 2016. *Plant Horror: Approaches to the Monstrous Vegetal in Fiction and Film*. London: Palgrave Macmillan.

Kerouac, Jack. 2007. *On the Road* [1957]. London: Penguin.

Khair, Tabish. 2009. *The Gothic, Postcolonislism and Otherness: Ghosts from Elsewhere*. London: Palgrave Macmillan.

Laskow, Sarah. 18 September 2018. "Nikola Tesla Built a Giant Tower to Send Wireless Electricity Around the World," *Atlas Obscura*. https://www.atlasobscura.com/articles/what-is-wardenclyffe-tower-nikola-tesla. Accessed 21 August 2019.

Le Fanu, Joseph Sheridan. 2013. *Carmilla: A Critical Edition*. Syracuse: Syracuse University Press.

Lines, Craig. 25 August 2015. "Corrado Farina interview: They Have Changed Their Face," *Den of Geek*. https://www.denofgeek.com/movies/corrado-farina/36530/corrado-farina-interview-they-have-changed-their-face. Accessed 21 August 2019.

Luckhurst, Roger. 2015. *Zombies: A Cultural History*. London: Reaktion Books.

Magistrale, Tony. 2008. *The Films of Stephen King: From Carrie to Secret Window*. London: Palgrave Macmillan.

Manolopoulos, Mark. 2009. *With Gifted Thinkers: Conversations with Caputo, Hart, Horner, Kearney, Keller, Rigby, Taylor, Wallace, Westphal.* Bern: Peter Lang.
Marryat, Florence. 2010. *Blood of the Vampire* [1897]. Brighton: Victorian Secrets.
Matheson, Richard. 2007. *I Am Legend* [1954]. London: Gollanz.
Matthewman, Sasha. 2011. *Teaching Secondary English as If the Planet Matters.* New York: Routledge.
Meehan, Paul. 2014. *The Vampire in Science Fiction Film and Literature.* Jefferson, NC: McFarland.
Melton, J. Gordon. 2011. *The Vampire Book: The Encyclopedia of the Undead.* Michigan: Visible Ink Press.
Melville, Herman. 2016. *Moby Dick* [1851]. London: Macmillan Collector's Library.
Mendik, Xavier. 2001. *Shocking Cinema of the Seventies.* Hereford: Noir Publishing.
Millar, Charlotte-Rose. 2019. *Witchcraft, the Devil, and Emotions in Early Modern England.* New York: Routledge.
Miller, Daniel. 5 June 2012. "Are you sure digging him up is a good idea? Archaeologists find Bulgarian 'vampires' from Middle Ages with iron rods staked through their chests," *Daily Mail.* https://www.dailymail.co.uk/sciencetech/article-2154837/Vampire-skeletons-unearthed-Bulgaria-iron-stakes-plunged-chests.html. Accessed 21st August 2019.
Miller, Elizabeth. 2012. *Dracula: Sense and Nonsense.* Southend-on-Sea: Desert Island Books.
Miller, Wayne. 31 May 2019. "More On That Book Proving Bram Stoker Knew About the Historical Dracula," *vampires.com.* https://www.vampires.com/more-on-that-book-proving-bram-stoker-knew-about-the-historical-dracula/. Accessed 21 August 2019.
Moretti, Franco. 1988. *Signs Taken for Wonders: Essays in the Sociology of Literary Forms,* trans. by Susan Fiscer, David Forgacs and David Miller. London: Verso.
Mulvey-Roberts, Marie. 2016. *Dangerous Bodies: Historicising the Gothic Corporeal.* Manchester: Manchester University Press.
Nadeau, Kathleen. December 2011. "Aswang and Other Kinds of Witches: A Comparative Analysis," *Philippine Quarterly of Culture and Society,* 39, 4, 250–66.
Nama, Adilifu. 2008. *Black Space: Imagining Race in Science Fiction Film.* Austin: University of Texas Press.
Nelson, Victoria. 2013. *Gothicka: Vampire Heroes, Human Gods, and the New Supernatural.* Cambridge: Harvard University Press.
Newsom, Carol A. 2009. *The Book of Job: A Contest of Moral Imaginations.* Oxford: Oxford University Press.
Niles, Steve, and Ben Templesmith. 2002. *30 Days of Night.* San Diego: IDW Publishing.
Page, Rachel E., Robert G. Weiner, and Cynthia J. Miller. 2012. "Billy the Kid vs. Dracula, about the strangest Western ever made without a pornographic subplot," in Cynthia J. Miller and A. Bowdoin Van Riper (eds.), *Undead in the West: Vampires, Zombies, Mummies, and Ghosts on the Cinematic Frontier.* Lanham: The Scarecrow Press Inc., 45–64.
Parker, Elizabeth. 2019. "Alex Garland's *Annihilation* (2018)—Eco-horror," in Simon Bacon (ed.), *Horror: A Companion.* Bern: Peter Lang Ltd., 93–102.
Peacock, M. Jess. 2015. *Such a Dark Thing: Theology of the Vampire Narrative in Popular Culture.* Eugene: Resource Publications.
Pheasant-Kelly, Frances. 2013. *Fantasy Film Post 9/11.* New York: Palgrave Macmillan.
Piatti-Farnell, Lorna. 2014. "Monsters of Capital: Vampires, Zombies, and Consumerism," in James Dow and Glen Whitman (eds.), *Economics of the Undead: Zombies, Vampires, and the Dismal Science.* Lanham: Rodman & Littlefield, 99–110.
Pollard, Tom. 2011. *Hollywood 9/11: Superheroes, Supervillains, and Super Disasters.* New York: Routledge.
Price, Cheryl Blake. 2013. "Vegetable Monsters: Man-Eating Trees in Fin-de-Siècle Fiction," *Victorian Literature and Culture,* 41, no. 2, 311–27.
Prosser, Ashleigh. 2018. "Joan Lindsay's *Picnic at Hanging Rock* (Book and Film)

(1967/1975)—"Australian Gothic," in Simon Bacon (ed.), *Gothic: A Reader*. Bern: Peter Lang, 87–96.
Radovic, Milja. 2017. *Film, Religion and Activist Citizens: An Ontology of Transformative Acts*. New York: Routledge.
Ramos, Maximo D. 1971. *The Aswang Complex in Philippine Folklore*. Scotts Valley: CreateSpace Independent Publishing Platform.
Ransom, Amy J. 2018. *I Am Legend As American Myth: Race and Masculinity in the Novel and Its Film Adaptations*. Jefferson, NC: McFarland.
Renzi, Thomas C. 2004. *H. G. Wells: Six Scientific Romances Adapted for Film*. Lanham: The Scarecrow Press Inc.
Rice-Burroughs, Edgar. 2012. *A Princess of Mars* [1912], in *Mars Trilogy*. New York: Saga Press, 3–226.
Robins, Nick. 2012. *The Corporation That Changed the World: How the East India Company Shaped the Modern Multinational*. London: Pluto Press.
Roy, Tirthankar. 2016. *The East India Company: The World's Most Powerful Corporation*. London: Penguin.
Schivelbusch, Wolfgang. 2014. *The Railway Journey: The Industrialization of Time and Space in the Nineteenth Century* [1977]. Oakland: University of California Press.
Senf, Carol A. 1988. *The Vampire in Nineteenth Century English Literature*. Madison: The University of Wisconsin Press.
Shelley, Mary. 2012. *Frankenstein* [1818]. New York City: W. W. Norton & Company.
Smith, Andrew and William Hughes (eds.). 2016. *Ecogothic*. Manchester: Manchester University Press.
Squires, Nick. 11 July 2010. "Dracula was not bloodthirsty, just a victim of bad propaganda, new exhibition claims," *The Telegraph*. https://www.telegraph.co.uk/news/worldnews/europe/romania/7883928/Dracula-was-not-bloodthirsty-just-a-victim-of-bad-propaganda-new-exhibition-claims.html. Accessed 21 August 2019.
Stoker, Bram. 1996. *Dracula* [1897]. London: Signet Classics.
_____. 2010. *The Lair of the White Worm* [1911], in *The Lair of the White Worm & The Lady of the Shroud*. London: Wordsworth Editions Ltd., 1–146.
Stone, Allucquère Rosanne. 1996. *The War of Desire and Technology at the Close of the Mechanical Age*. Cambridge: MIT Press.
Summers, Montague. 2013. *The Vampire in Europe*. New York: Routledge.
_____. 2017. *The Vampire, His Kith and Kin* [1929]. Bristol: Mockingbird Press.
Sykes, Brad. 2018. *Terror in the Desert: Dark Cinema of the American Southwest*. Jefferson, NC: McFarland.
Taunton, Matthew. 15 May 2014. "Class in The Time Machine," *British Library*, Discovering Literature: Romantics & Victorians. https://www.bl.uk/romantics-and-victorians/articles/class-in-the-time-machine. Accessed 21 August 2019.
Traina, Vincenzo. 1980. "Artificial Insemination and Semen Banks in Italy," in Georges David and Wendell S. Price (eds.), *Human Artificial Insemination and Semen Preservation*. Boston: Springer.
Various authors. 1938–1956. *Lebor Gabála Érenn*, a.k.a. The Book of the Taking of Ireland, a.k.a. The Book of Invasion [c. 1150], trans. by Robert Alexander Stewart Macalister. Dublin: Irish Texts Society.
Voltaire. 1927. *The Works of Voltaire, a Contemporary Version; a Critique and Biography-Edition Deluxe-Volume VII-Philosophical Dictionary*, trans. by William F. Fleming. New York: Dingwall Rock Ltd.
Wald, Priscilla. 2008. *Contagious: Cultures, Carriers, and the Outbreak Narrative*. Durham: Duke University Press.
Weber, Max. 2010. *The Protestant Ethic and the Spirit of Capitalism* [1920], trans. by Stephen Kalberg. Oxford: Oxford University Press.

Wells, H. G. 2014. "The Flowering of the Strange Orchid" [1894]. Scotts Valley: Create Space Independent Publishing Platform.
_____. 2016. "The Red Room" [1894], in *The Red Room & Other Horrors: H. G. Wells' Best Weird Science Fiction and Ghost Stories, Annotated and Illustrated*. Fort Wayne: Old Style Tales, 21–31.
_____. 2018. *War of the Worlds* [1897]. London: Penguin.
Weston, Phoebe. 22 June 2019. "The rise of eco-anxiety and how to come to terms with climate change," *Independent*. https://www.independent.co.uk/environment/climate-change-extinction-eco-anxiety-ice-melting-sea-level-wildfire-a8968011.html. Accessed 21 August 2019.
Wicke, Jennifer. Summer, 1992. "Vampiric Typewriting: Dracula and Its Media," *ELH* 59/2, 467–93.
Wilberg, Peter. 2009. *Event Horizon: Terror, Tantra and the Ultimate Metaphysics of Awareness*. Scotts Valley: CreateSpace Independent Publishing Platform.
Wyndham, John. 2008. *The Day of the Triffids* [1951]. London: Penguin.
Yarboro, Chelsea Quin. 2014. *Hôtel Transylvania: A Timeless Novel of Love and Peril* [1978]. New York: Open Road.
Young-Roberts, Bryn V. 2012. *The Film Reader's Guide to James Cameron's Avatar*. Morrisville: Lulu.com.

Index

abyss 6, 9, 25, 33, 34, 114, 162, 191–2
Africa 56, 93, 94, 96, 119, 120
African American 29, 118–9, 122, 124, 125, 160
alien 11, 14, 20, 27, 36, 43, 60, 76, 85, 107, 108, 109, 111–2, 141, 142, 154, 155–6, 158, 160, 161, 164, 169–73, 175, 177, 178, 180, 183, 185–7, 189
America 9, 27, 29, 32, 36, 40, 50, 72, 88, 89, 100, 101–2, 103, 113, 118, 120–1, 124, 126, 132, 137, 143, 158, 173, 183, 186, 188, 189
animal 3, 4, 8, 10, 12, 35, 42, 51–3, 60, 74, 76, 79, 80, 84, 86, 89, 130, 160, 161, 170, 173, 184, 192
animalism 16, 27, 38, 109
anthropocene 1, 113, 152, 191
antibody 49, 57
apocalypse 10, 33, 36, 38, 53, 55, 83, 86–7, 89, 94, 101, 106, 122–3, 160, 165, 188; post 102, 114, 126, 136
autoimmune 107, 111
automobile 98, 126, 131

bacteria 49, 107, 157
bats 9, 16, 21, 27, 32, 42, 48–53, 55, 57, 71, 87–8, 91, 102, 114, 192
biological 7, 12, 23, 25, 42, 47, 49–50, 58, 60–1, 84, 88, 94, 111, 154, 167, 170–4, 180, 182; weapon 8–9, 101, 121
bioweapon 25, 122, 135,
bite 3, 37, 49, 50, 73, 81, 86, 99
blood 2, 3, 5, 9, 11, 15, 18–19, 23, 29, 31, 36, 40, 45–6, 48–50, 53, 58, 61–2, 65, 67, 71, 73, 75, 76, 78, 84–5, 87, 93–5, 98–101, 108–10, 120–1, 125, 128, 120–32, 137–8, 140, 144, 155–6, 165–8, 172, 182; sucking 7, 89, 122, 147

cannibalism 108, 109, 137, 138, 139
capitalism 4, 5, 13, 17, 18, 19, 21, 55, 106, 119, 122, 124, 125, 132–4, 137, 140, 143, 146, 160, 162, 190
Carmilla 3, 128
castle 3, 13–7, 19–20, 23, 25–6, 31, 42–3, 45–6, 48, 54, 70–1, 132, 147–8, 165
chaos 25, 93, 95, 111, 170, 182
child 3, 18, 38, 53, 64, 66, 76, 78, 81, 91, 92, 107, 114–6, 125, 131, 133, 135, 136, 138–9, 145, 152, 186

city 28, 32, 33, 40, 41, 86, 89, 95, 98–9, 101, 111–3, 123–5, 126, 133, 141, 148, 150–1, 170, 183, 185
civilization 5, 30, 32, 37, 39, 73, 75, 82, 84–5, 87–9, 99–102, 104, 107, 111–3, 116, 121–2, 132, 133, 143, 146, 149, 170
climate 2, 7, 33, 108, 131, 136, 144, 190
colonial 2, 3, 6, 7, 14, 19, 55, 72, 97, 101, 119, 143, 144, 155, 157, 158, 159, 162, 165, 192; eco 79, 154; post 73; reverse 4, 6, 154
consumerism 1, 4, 5, 8–9, 11, 12, 15, 17, 18, 25, 45, 72, 90, 93, 98, 99, 106, 111, 113, 118–23, 125–6, 128–32, 134, 135, 139, 142–5, 146, 148, 150, 151, 153, 156, 161, 162, 179, 186, 189–90, 192; non 96; post 146, 183
contagion 10, 12, 23–4, 49, 85, 97, 99, 102–3, 108, 110, 122, 162, 165, 173, 174
Count Dracula 1, 2, 4, 7, 9, 14, 15–9, 23–6, 28, 31, 43, 45, 48, 49, 54, 61, 63, 68, 69, 72, 109, 112, 123, 126, 132, 143–5, 147, 156, 165, 176, 189
creature 1–3, 5, 6, 8, 9, 10, 16, 22, 24, 31, 34, 38, 41–2, 44, 46, 48, 50, 56–7, 60–1, 67, 68, 71, 73–5, 81–2, 84, 85–6, 100, 102, 107, 109, 111–3, 142, 156, 163–4, 166, 169, 171, 172–5, 176; feature
curse 23, 90

dark: ecology 53, 77–8, 85, 162, 180; force 10, 13, 20, 21, 41, 71, 76, 79, 105, 132, 143; mirror 8
darkness 23, 24, 26, 29, 34, 37, 39, 56, 70, 108, 124, 163, 164, 165
death 1, 3, 4, 14, 20, 21, 23–25, 33, 35, 37, 40, 45, 51, 52, 71, 80, 90, 95, 107, 111, 125, 127, 130, 131, 137, 142, 149, 152, 157, 162, 168, 182, 184, 189
degeneration 4, 12
demonic 1, 29, 48, 71, 126, 133
desert 9, 26, 27–32, 34, 35, 149, 150, 151–2
Devil 41, 69, 75
disease 5, 10, 45, 50, 84–5, 88, 107, 108, 114, 155, 157
Dracula, novel 3, 5, 7, 11, 18, 19, 46, 48, 52, 68, 76, 82, 118, 121, 123, 132, 133, 146, 163, 176

205

Index

eco: activism 4, 36; colonialism 51, 79; credentials 131; friendly 1, 2, 11, 131; gothic 4, 6, 7; horror 4–7, 77, 112; immune 162; phobic 44; savior 143; system 1–6, 8, 9–12, 15, 16, 25, 27, 29, 31, 35–6, 42, 45–6, 48, 52, 57, 61, 64, 70, 72, 75, 82, 83, 92, 99, 100, 107, 110, 114, 116–7, 118, 129, 137, 141, 148, 149, 152, 154–7, 159, 167, 171–2, 174, 182, 187, 191–2; threat 52; topian 104, 106; vampirism 26, 47; warrior 1, 3, 50, 114, 192
ecological 5, 6, 7, 9, 13, 18–9, 21, 23, 33, 37, 40, 50, 55, 62, 76, 84–6, 88–9, 93, 96–7, 103, 106, 120–2, 126, 130, 134, 140, 146, 153, 158, 162–3, 170, 178, 181, 186,; anti 112, 183
economics 12, 101, 116, 125, 143, 145, 146, 158, 165, 176, 183, 186, 192,
electricity 14, 18, 144, 145, 172, 190
England 4, 14, 16, 18, 19, 76, 170
energy 23, 25, 47, 71, 75, 79, 88, 92, 94, 117, 129, 140, 142, 143, 145, 147, 167, 169, 170, 177; bio 60, 72; vampire 1, 3, 7, 62, 63, 64, 68, 78, 107, 128, 141, 155, 165, 170, 175, 176, 192
environment 2–6, 7, 9–10, 14–6, 18, 20, 22, 25–7, 32, 35–7, 39, 42–3, 48–9, 50–3, 55, 60, 61–2, 64, 68, 71–3, 75–6, 78, 79, 82, 84, 86, 88–9, 92–4, 96, 97–8, 100, 102–3, 106, 110, 11–4, 116, 120, 124, 129, 131, 133, 137, 140, 142, 144, 146, 156–62, 167, 170, 176, 183, 185, 187, 191
ethnicity 13, 120
evil 6, 14, 21, 52, 53, 55, 56, 70, 77–8, 123, 182
evolution 10, 23, 38, 39, 40, 42, 48, 56, 85, 88, 97, 111, 116, 132, 146, 147, 150, 152, 155–6, 171, 192
existential 52, 53, 61, 68, 72, 75, 88, 98, 107, 119, 121, 139, 140, 144, 147, 150, 152, 156, 165, 172, 174, 192
extinction 1, 4, 88, 136, 149, 150, 152, 154, 157, 166
extraterrestrial 12, 126, 162, 165, 183, 188

female 2, 19, 41, 45, 53, 75, 95, 147, 151, 168–70
feminine 5, 128, 169
feral 35, 99, 100, 109, 133
flesh 14, 60, 82, 96, 166, 173; eating 35, 94, 106, 108, 113, 114, 116, 133
flower 7, 24, 60, 64, 81, 82, 120
folklore 1, 10, 49, 58, 85, 86, 110, 171, 191
forest 5, 9, 11, 38, 40, 56, 57, 58, 66, 68, 76–81, 102, 113–4, 116, 149, 160, 165, 188, 191
fuel 11, 18, 28, 34, 52, 58, 71, 94, 126, 128–31, 133, 143, 144–6
fungus 61, 77, 79, 80–2, 114–6, 192
future 5, 10–11, 18, 19, 20, 25, 27, 28, 35, 37, 46, 49, 51, 70–1, 72, 83, 86–8, 97, 100, 101, 103–4, 106–7, 112, 114, 116, 121, 129–32, 136, 139, 144, 146, 149, 150, 158, 165, 168, 175, 179, 180, 183, 192

gaia 107, 161
global 11, 41, 51, 88, 94, 100, 107, 131, 132, 136, 153, 158, 174, 190

gothic 4, 5, 6, 10, 72, 88, 108, 132, 133, 145, 173, 180
greed 39, 50, 51, 73, 92, 131, 132, 154, 178
ground 9, 14, 41, 42, 52–4, 62, 63–4, 66, 73, 78, 81, 89, 104, 106, 110, 115, 119, 120, 134, 144

habitat 4, 9, 39, 44, 50–3, 79, 116
Harker, Jonathan 13–20, 24–5, 28, 42–3, 45, 63, 70, 73, 133
history 3, 13, 15, 17, 27, 29, 36, 39, 52, 54–6, 61–2, 69–70, 79, 93, 95, 101, 106, 109, 112, 113, 119, 120, 132, 141, 144, 191
home 1, 3, 4, 10, 14–5, 17–9, 21, 23, 32, 33, 37, 50, 51, 57, 68, 70, 72, 76, 80, 87, 90, 94, 97, 98, 102, 112, 119, 123, 129, 132, 135, 141–2, 146, 148, 156, 158, 161, 163, 165–8, 178, 185, 187, 188, 192
horror 6, 10, 21, 29, 33, 50, 108, 137, 170, 173
hubris 84, 179
humankind 1, 8, 55, 82, 85, 107, 115, 119, 149, 174, 175
hybrid 42, 43, 44, 49, 88, 97, 99, 101, 103, 115–6, 121, 142, 147, 150, 156, 185, 189

I Am Legend (novel) 5, 10, 49, 83, 86–8, 95, 98, 101, 103, 121, 136
immunity 49, 84, 87, 104, 107, 114, 122, 161, 162, 167, 180
imperialism 18, 19, 43, 79, 88, 154, 155, 162
indigenous 14, 27, 32, 33, 36, 52, 57, 101, 154, 155, 159, 160
industrialization 4, 5, 6, 8, 10, 13, 16, 18, 101, 118, 132, 133, 144, 145
insatiable 51, 87, 118, 138, 164–5
insects 1, 21, 50, 58, 73, 114, 138, 152, 189, 192
invasion 11, 16, 50, 79, 154–5, 156, 158, 167, 169

jouissance 7, 12, 43, 58, 60, 112, 170, 171, 182

kingdom 70–2

lair 20, 28, 42, 54–5, 102, 108, 132, 133–4, 146–8
land beyond the forest 5, 11, 13, 25, 43, 70, 146, 188
landscape 2, 4, 5, 8, 9, 11, 13–5, 19, 20, 25, 27, 28, 30, 32, 33, 39, 41, 55, 61, 71, 72, 76, 89, 97, 109, 149, 162, 163, 183, 189, 191
London 6, 20, 93, 95, 103, 104, 115, 118, 143, 168, 170
Lucy (character) 17, 18, 19, 23, 26, 48

machine 11, 12, 17, 99, 103–6, 122, 123, 126, 130, 132, 148, 152, 156, 159, 160, 169, 182, 189, 190
male 5, 11, 45, 57, 92, 95, 105, 141, 168, 169, 170
mankind 1, 3, 7, 10, 11, 21, 51, 83, 84, 89, 92, 94, 104, 107, 111, 114, 116, 126, 152, 157, 179, 181, 191, 192
manmade 49, 72, 102, 178, 189
masculine 5, 11, 123, 128

Index

metropolis 20, 28, 89, 98, 111, 122, 150
military 88, 98, 115, 122, 144, 158–61, 185, 188–9
Mina (character) 17, 18, 19, 23, 24, 63, 170
mirror 8, 11, 39, 46, 70, 84, 85, 86, 107, 109, 137, 154, 156, 158, 162, 165, 178, 180
modernism 6, 8, 9, 13, 15, 18, 19, 24, 27, 53, 55, 64, 79, 102, 120, 126, 132, 134, 192
monster 2, 4, 5, 44, 45, 70, 71, 73, 94, 96, 110, 111, 114, 116, 121, 124, 145, 163–5, 171, 172–4
monstrous 6, 10, 39, 55, 72, 74, 101
Mother 3, 39, 56, 62–4, 68, 71, 78, 96, 115, 183, 185; Nature 52, 149
mutation 12, 42, 43, 44–5, 49, 50–1, 57, 59, 60, 74, 84, 86, 88, 102, 116, 142, 174, 185
myth 27, 31, 56, 79, 90, 120, 140, 141, 185

Nature 1, 2, 5–6, 7–9, 11, 12, 13, 14, 16, 20–1, 23, 24–7, 29, 40, 44, 49, 51–2, 55, 58, 70, 79, 83, 85, 90, 94, 100, 111, 114, 133–4, 149, 162, 173, 180, 183, 191–2
neoliberal 123, 125, 131, 137, 192
Nosferatu (Murnau: 1922) 6, 8, 11, 21–4, 29, 36, 43, 52, 80, 128, 163, 178, 181

ocean 16, 154, 176
organic 7, 10, 17, 20, 23, 36, 43, 44–6, 60, 61, 75, 76, 134, 142, 146, 162, 167, 169, 180–1, 189
organism 103, 107, 116, 172
Orlok (character) 9, 12, 20–4, 36, 40, 43, 45, 52, 109, 128, 141, 142, 163, 178, 189
outbreak 85, 86, 89, 94, 99, 102, 114, 174, 191; post 98
outer space 1, 11, 12, 42, 107, 141, 154, 161, 162, 165, 178, 188

parasite 5, 141, 167, 185
patriarchy 2, 29
pestilence 23, 174, 187, 189–90
plague 25, 26, 49, 52, 82, 83, 84, 86–8, 97, 102, 109, 111, 121–5, 149, 184, 187, 189–90, 192
plant 6–7, 10, 11, 43, 45, 55, 62, 73, 89, 90, 110, 138, 157; carnivorous 21, 91–2, 93–7; life form 36, 44, 168, 172, 175
predator 12, 21, 22, 24, 25, 2, 44, 75
pregnant 3, 56, 57, 74, 115

rabid 110, 170
rainforest 9, 113, 114, 116, 165
rats 3, 9, 16, 21, 23, 24–5, 26, 52
reproduction 7, 57, 75, 114, 119, 172; anti 135
resources 6, 9, 11, 16, 18, 19, 26, 27, 34, 36, 70, 76, 79, 101, 113, 126, 132, 143, 149, 154–5, 158–9, 162, 176, 184, 188, 192
revenge 31, 82, 84, 88, 104, 107, 128, 145, 146, 160, 191

robot 11, 46, 149–52, 159, 161, 167, 186, 188–9
rural 9, 13, 20, 32, 40, 42, 96, 101, 192

sacrifice 9, 23, 67–8, 78, 93, 106, 109, 119–20, 125, 137, 138, 139
Satan 41, 53, 76, 77, 78
self: consuming 8, 111, 153; regulation 107, 147, 148; sustaining 10, 11, 106, 137
skin 63, 70, 87, 88, 93, 99, 125, 128, 141, 147, 173, 184, 187
soil 3, 15, 19, 23, 74, 77, 149, 156
South America 2, 9, 48, 50, 52, 113, 114, 144
spectral 20, 192
sublime 5, 8, 9, 13, 25, 26, 28, 40, 42, 43
supernatural 1, 2, 11, 13, 26, 40, 48, 57, 58, 72, 84, 86, 96, 109, 142–5, 163, 165, 185
superstition 13, 85, 132, 191

technology 4, 8, 9, 13, 17, 18, 20, 27, 45, 94, 102, 121, 122, 123, 125, 140–3, 144, 145–6, 149, 151, 155–6, 158–61, 171, 180, 187, 188
toxic 50–1, 190
Transylvania 4, 9, 13, 14, 18–20, 27, 28, 46, 61, 69–70, 73, 113, 133
trauma 27, 34, 36, 39, 55, 61, 62, 85, 86, 93, 96, 101, 165

undead 1–4, 5, 7, 8, 24, 33, 34, 37, 39, 48, 53, 61, 62, 68, 86, 98, 99, 101, 103, 109, 110, 115, 121, 124, 140, 143, 145, 167, 168, 191
urban 13, 33, 40, 41, 42, 96, 133, 192

Van Helsing 16, 17, 19, 21, 26, 52, 55, 123, 144–6
vegetarian 2
vegetation 6, 10, 77, 89, 93, 104, 115, 123, 156
Venus flytrap 6, 12, 21, 64, 178
virus 60, 85, 86, 87, 88, 108, 109, 110

War of the Worlds (novel) 3, 11, 51, 83, 85, 107, 154, 155, 157, 159, 162, 167
wealth 51, 72, 90, 98, 119, 125, 131, 132, 136, 137, 143, 150
weather 8, 16, 28, 34, 63, 72, 102, 109, 113
whiteness 11, 20, 29, 72, 73, 92, 119, 120–2, 124–5, 140, 141, 159, 160, 187
witch 2, 41, 56, 75–9, 90, 92, 143
wolf 16, 19, 27, 133, 142, 185; were 2, 20, 42
womb 45, 56
woodlands 9, 10, 40, 41, 42, 52, 61, 68, 71, 76, 77, 79, 80, 81, 82, 101

xenomorph 45, 155

zombie 5, 33, 60, 84, 86, 90, 94–6, 101, 102, 114, 115, 170

www.ingramcontent.com/pod-product-compliance
Ingram Content Group UK Ltd.
Pitfield, Milton Keynes, MK11 3LW, UK
UKHW042002140426
5217IPUK00015B/935